E. Layer

Modelling of Simplified Dynamical Systems

Springer

*Berlin
Heidelberg
New York
Barcelona
Hong Kong
London
Milan
Paris
Tokyo*

Edward Layer

Modelling of Simplified Dynamical Systems

With 55 Figures

 Springer

Edward Layer DSc, PhD

Cracow University of Technology
Faculty of Electrical and Computer Engineering
Institute of Electrical Metrology
24 Warszawska Str.
31-155 Krakow
Poland

ISBN 3-540-43762-2 Springer-Verlag Berlin Heidelberg New York

Library of Congress Cataloging-in-Publication Data
Layer, Edward, 1942-
Modelling of simplified dynamical systems / Edward Layer.
   p. cm.
Includes bibliographical references and index.
   ISBN 3540437622
   1. Mathematical models. I. Title.
QA401 .L39 2002   511'.8--dc21   2002075897

Springer-Verlag Berlin Heidelberg New York
a member of BertelsmannSpringer Science+Business Media GmbH

http://www.springer.de

© Springer-Verlag Berlin Heidelberg 2002
Printed in Germany

Typesetting: Dataconversion by author
Cover-design: de'blik, Berlin
Printed on acid-free paper   SPIN: 10879558   62 / 3020 hu - 5 4 3 2 1 0 -

TABLE OF CONTENTS

1. INTRODUCTION

Problems involving synthesis of mathematical models of various physical systems, making use of these models in practice and verifying them qualitatively has become an especially important area of research since more and more physical experiments are being replaced by computer simulations. Such simulations should make it possible to carry out a comprehensive analysis of the various properties of the system being modelled. Most importantly its dynamic properties can be addressed in a situation where this would be difficult or even impossible to achieve through a direct physical experiment. To carry out a simulation of a real, physically existing system it is necessary to have its mathematical description; the system being described mathematically by equations, which include certain variables, their derivatives and integrals. If a single independent variable is sufficient in order to describe the system, then derivatives and integrals with respect to only that variable will appear in the equations. Differentiation of the equation allows the integrals to be eliminated and produces an equation which includes derivatives with respect to only one independent variable i.e. an ordinary differential equation. In practice, most physical systems can be described with sufficient accuracy by linear differential equations with time invariant coefficients. Chapter 2 is devoted to the description of models by such equations, with time as the independent variable. Other forms of model description with transfer function and state equations are also presented in this chapter, since they are very popular in electrical and control engineering. Additionally, methods of solving the respective equations, which will repeatedly be used in this work, are discussed.

It is obvious that the success of any simulation is dependent mainly on the validity of the model that represents the system being investigated. It is impossible to achieve one hundred percent accuracy and any attempt to approach it requires more and more complicated models to be built. In consequence, any increase in model validity requires the application of equations of higher and higher order, often resulting in a higher degree of complication. However, the time consuming and costly process of analysing their properties often outweighs the practical advantages resulting from the application of these complicated models. Therefore models of lower order are used in many cases, accepting their reduced accuracy, even if a more complicated model of higher order is known. Chapter 3 presents the basic parameters which characterise the dynamic properties of systems, especially those of low orders, while Chapters 4 and 5 are devoted to the discussion of methods of model synthesis and the various ways they can be simplified. Problems of model simplification are directly related to the necessity of determining the error with which a reduced model maps the primary model. Methods of determining such errors for different objective functions are presented in literature, however they are generally stated for standard input signals and among them the unit step input is most often encountered. Nevertheless the error values determined for standard input signals are of little use, because in many cases in practice, signals which could operate on the system being modelled differ considerably from standard signals. Furthermore, we often have to deal with systems, e.g. measuring systems, for which the input signals are indeterminate of arbitrary shape, being impossible to be foreseen a priori, as for instance signals originating from seismic waves, signals of acoustic pressure generated by explosions of explosive charges, signals from earth crust vibrations caused by earthquakes, etc. As it is impossible to analyse the full set of all imaginable input dynamic signals, it is suggested to solve this problem by using the maximum values of assumed error criteria, since the mapping error values being determined should be credible for any input. However, determination of maximum error requires the application of special input signals that can often be difficult for analytical determination and in many cases can be only found by means of computer simulation. Problems concerning the determination of maximum error values and of the signals necessary are treated in Chapter 6. This section presents both the analytical methods as well as the necessary algorithms that enable signals to be determined which maximise the integral-square-error criterion through computer simulation. Also demonstrated is a way of calculating its value over a given time interval. Signals with a single, as well as with two simultaneous constraints are considered here, which permits their dynamic behaviour to be matched to the dynamics of the models under testing. Chapter 7 presents the practical application of the theory developed in Chapter 6 and explains the effect of the resulting algorithms on the example of the *minimax* type optimisation of a selected class of filters. This is performed with the aid of signals with single and double constraints and of integral-square-error criterion.

The book is devoted to selected problems concerning investigations into dynamic model properties, methods of simplifying high-order models, determining mapping errors of simplified models, their optimisation and the synthesis of input signals being suitable for this purpose. Examples of calculations included have been selected in such a way so as to illustrate the theory being presented. They are often the result of numerous trials, but this does not mean that similar problems, which may be encountered by the reader in practice, can be solved in an identical way.

The Author believes that this book will be useful both for students of engineering departments, as well as for engineers and scientists in various fields of technology, who in their work, make use of modelling and computer simulation. He also hopes that the content of this book will not present the reader with difficulty in understanding it and that it will provide the basis for a possible follow-up.

2. MATHEMATICAL MODELS

A precondition, which permits any computer simulation of a physically existing system to be carried out, is the creation of its mathematical model. Such a model is most often obtained as a result of identification of the system or less frequently on the basis of its structural analysis, if possible. However not all models are obtained in this way. Therefore it is worth noting that within the wide set of mathematical models of various physical systems, a certain small subset can be separated. This subset is characterised by the fact that for its models there are legally and officially testified values of parameters, defined most often by their rated data. Those models are usually relatively simple and their form can easily be determined. This form results from an objective function that is an abstract mathematical formula, which should be met by the system to be modelled. Therefore such models are called models of standards. Constituting a reference, models of standards play an important role in determining errors, mainly in automatic control systems, dynamic metrology, etc.

Physically existing systems are non-linear, but in most cases their non-linearity is small enough so that the errors caused by it can be ignored. Accordingly, we will describe them in the time domain by linear differential equations or by state equations and in the domain of the Laplace transform by transfer functions. Following this, these methods of description will be discussed, along with methods of solving the corresponding equations and finding the mutual relationships occurring between them. In the conclusion of the chapter we will discuss selected models of standards and their properties.

2.1. Differential equations

A single input time invariant linear dynamic model, described by means of a non-homogeneous differential equation of the n–th order, has the following form for zero initial conditions

$$\frac{d^n y(t)}{dt^n} + a_{n-1}\frac{d^{n-1}y(t)}{dt^{n-1}} + ... + a_1\frac{dy(t)}{dt} + a_0 y(t)$$

$$= b_0 u(t) + b_1\frac{du(t)}{dt} + ... + b_{m-1}\frac{d^{m-1}u(t)}{dt^{m-1}} + b_m\frac{d^m u(t)}{dt^m} \tag{2.1}$$

$$y(0) = 0, \ y'(0) = 0, \ ... \ y^{(n-1)}(0) = 0 \quad m < n, \quad a_k, b_k \in \Re$$

where $u(t)$ is input function and $y(t)$ is output function.

The solution of the non-homogeneous differential equation (2.1) is the sum of the homogeneous solution called the general integral and the particular solution sometime also called a particular integral. The homogeneous solution

$$(D^n + a_{n-1}D^{n-1} + ... + a_1 D + a_0)y(t) = 0 \tag{2.2}$$

where

$$D^k y(t) = \frac{d^k y(t)}{dt^k} \tag{2.3}$$

is sought in the form

$$y(t) = e^{st} \tag{2.4}$$

where s is the root of the characteristic equation

$$D^n + a_{n-1}D^{n-1} + ... + a_1 D + a_0 = 0 . \tag{2.5}$$

The four following cases can occur for equation (2.5) [51], [84], [85]

- If the characteristic equation has n different real roots s_1, s_2, ... s_n then the solution $y(t)$ is

$$y(t) = c_1 e^{s_1 t} + c_2 e^{s_2 t} + ... + c_n e^{s_n t} . \tag{2.6}$$

- If the characteristic equation has n different roots among which there are k conjugate complex roots: $s_1 = \alpha_1 + j\beta_1$, $s_2 = \alpha_2 + j\beta_2$, ... $s_k = \alpha_k + j\beta_k$, $s_{k+1} = \alpha_1 - j\beta_1$, $s_{k+2} = \alpha_2 - j\beta_2$, ... $s_{2k} = \alpha_k - j\beta_k$, then

$$y(t) = c_1 e^{\alpha_1 t} \sin \beta_1 t + ... + c_k e^{\alpha_k t} \sin \beta_k t + c_{k+1} e^{\alpha_1 t} \cos \beta_1 t + ... \tag{2.7}$$
$$+ c_{2k} e^{\alpha_k t} \cos \beta_k t + ... + c_{2k+1} e^{s_{k+1} t} + ... + c_n e^{s_n t} .$$

- If the characteristic equation has n real roots, among which there are k multiple roots $s_1 = s_2 = ... s_k$ and the remaining solutions are simple roots $s_{k+1} \neq s_{k+2} \neq ... s_n$, then

$$y(t) = (c_1 + c_2 t + ... + c_k t^{k-1}) e^{s_k t} + c_{k+1} e^{s_{k+1} t} + ... + c_n e^{s_n t}. \tag{2.8}$$

- If the characteristic equation has n roots, among which there are k complex conjugate roots $s_1 = s_2 = ... s_k = \alpha_1 + j\beta_1$, $\quad s_{k+1} = s_{k+2} = ... s_{2k} = \alpha_1 - j\beta_1$, then

$$y(t) = (c_1 + c_2 t + ... + c_k t^{k-1}) e^{\alpha_1 t} \sin \beta_1 t$$
$$+ (c_{k+1} + c_{k+2} t + ... + c_{2k} t^{k-1}) e^{\alpha_1 t} \cdot \cos \beta_1 t + c_{2k+1} e^{s_{2k+1} t} + ... + c_n e^{s_n t}. \tag{2.9}$$

When the characteristic equation has two or more pairs of multiple complex roots, the solution to the homogeneous equation is sought in a similar way.

The particular solution to the non-homogeneous equation can be conveniently determined by the indeterminate coefficients method. This method depends on finding the homogeneous solution by using the annihilator of the right hand side of the non-homogeneous equation. Let us denote the right hand side of the equation (2.1) as $g(t)$

$$g(t) = b_0 u(t) + b_1 \frac{du(t)}{dt} + ... + b_{m-1} \frac{d^{m-1} u(t)}{dt^{m-1}} + b_m \frac{d^m u(t)}{dt^m}. \tag{2.10}$$

Taking into consideration (2.2) this becomes

$$(D^{(n)} + a_{n-1} D^{(n-1)} + ... + a_1 D + a_0) y(t) = g(t). \tag{2.11}$$

The polynomial W of the differentiation operator D

$$W(D) = D^{(n)} + a_{n-1} D^{(n-1)} + ... + a_1 D + a_0 \tag{2.12}$$

represents annihilator $y(t)$ if

$$W(D) y(t) = 0. \tag{2.13}$$

Introducing the relationships (2.10) - (2.12) into (2.1) we can write

$$W(D) y(t) = g(t). \tag{2.14}$$

In the case of $g(t)$ being

$$\left\{\begin{array}{l} \text{constant } a \\ \text{polynomial } t \\ \text{exponential function } e^{\alpha t} \\ \sin\beta t, \ \cos\beta t \end{array}\right\} \tag{2.15}$$

or a ring of the above functions, there can be always found such an annihilator $W_1(D)$ for which

$$W_1(D)g(t) = 0. \tag{2.16}$$

Substituting formula (2.14) into (2.16) we obtain

$$W_1(D)W(D)y(t) = W_1(D)g(t) = 0. \tag{2.17}$$

In the same way, we can solve the homogeneous equation

$$W_1(D)W(D)y(t) = 0 \tag{2.18}$$

and obtain the general solution of non–homogeneous equation (2.1). Based on this solution, it is possible to find the particular integral to this equation. It is easy to verify that the annihilators $W_1(D)$ for the functions given by formula (2.15) can be of the following form:

$$D^n \left\{\begin{array}{l} 1 \\ t \\ t^2 \\ \cdot \\ t^{n-1} \end{array}\right\} = 0. \tag{2.19}$$

$$(D-\alpha)^n \left\{\begin{array}{l} e^{\alpha t} \\ te^{\alpha t} \\ t^2 e^{\alpha t} \\ \cdot \\ t^{n-1}e^{\alpha t} \end{array}\right\} = 0. \tag{2.20}$$

$$[D^2 - 2\alpha D + (\alpha^2 + \beta^2)]^n \left\{ \begin{array}{c} e^{\alpha t}\cos\beta t \\ te^{\alpha t}\cos\beta t \\ t^2 e^{\alpha t}\cos\beta t \\ \cdot \\ t^{n-1}e^{\alpha t}\cos\beta t \end{array} \right\} = 0 . \tag{2.21}$$

$$[D^2 - 2\alpha D + (\alpha^2 + \beta^2)]^n \left\{ \begin{array}{c} e^{\alpha t}\sin\beta t \\ te^{\alpha t}\sin\beta t \\ t^2 e^{\alpha t}\sin\beta t \\ \cdot \\ t^{n-1}e^{\alpha t}\sin\beta t \end{array} \right\} = 0 . \tag{2.22}$$

In the case when $\alpha = 0$ and $n = 1$ formulae (2.21) and (2.22) assume a simpler form

$$(D^2 + \beta^2)\left\{ \begin{array}{c} \cos\beta t \\ \sin\beta t \end{array} \right\} = 0 \tag{2.23}$$

which can be often encountered in practice.

2.2. Transfer function

If we apply the Laplace transform to both sides of equation (2.1) we obtain a description of the model by means of transfer function, which is a practice commonly used in electrical and control engineering.

$$K(s) = \frac{Y(s)}{U(s)} = \frac{b_m s^m + b_{m-1}s^{m-1} + ... + b_1 s + b_0}{s^n + a_{n-1}s^{n-1} + ...a_1 s + a_0} \tag{2.24}$$

where

$$Y(s) = \int_0^\infty y(t)e^{-st}dt, \;\; U(s) = \int_0^\infty u(t)e^{-st}dt, \;\; \frac{d^k y(t)}{dt^k} = s^k Y(s), \frac{d^k u(t)}{dt^k} = s^k Y(s)$$

In the case where a transform of the input function $U(s)$ is given, the solution $y(t)$ of equation (2.24) can be easily obtained by making use of the residue method. Two cases can occur here:

- The model has n single poles s_i, then

$$y(t) = \sum_{i=1}^{n} \operatorname{res} Y(s) e^{s_i t} \qquad s = 1, 2, \ldots n \tag{2.25}$$

where in (2.25)

$$\operatorname{res} Y(s) = \lim_{s \to s_i} (s - s_i) Y(s). \tag{2.26}$$

- The model has multiple poles s_m, then

$$y(t) = \sum_{k=1}^{r} \operatorname{res} Y(s) \frac{t^{(k-1)}}{(k-1)!} e^{s_m t} \qquad k = 1, 2, \ldots r \tag{2.27}$$

where in (2.27)

$$\operatorname{res} Y(s) = \frac{1}{(r-k)!} \lim_{s \to s_m} \frac{d^{(r-k)}}{ds^{(r-k)}} [(s - s_m)^r Y(s)] \qquad k = 1, 2, \ldots r \tag{2.28}$$

and r is the order of the multiple pole s_m.

In the case where single and multiple poles appear simultaneously, the solution of equation (2.24) is given by the sum of (2.25) and (2.27)

$$y(t) = \sum_{i=1}^{n} \operatorname{res} Y(s) e^{s_i t} + \sum_{k=1}^{r} \operatorname{res} Y(s) \frac{t^{(k-1)}}{(k-1)!} e^{s_m t}. \tag{2.29}$$

2.3. State equations

Differential equation can be written in the form of state equation

$$\begin{aligned} \dot{x}(t) &= \mathbf{A}x(t) + \mathbf{B}u(t), \quad x(0) = 0 \\ y(t) &= \mathbf{C}x(t) \end{aligned} \tag{2.30}$$

where $\mathbf{x}(t)$ is state vector, $\mathbf{A}$, $\mathbf{B}$ and $\mathbf{C}$ are real matrices of corresponding dimensions.

For single input $u(t)$ and single output $y(t)$ (2.30) assumes a simpler form

$$\dot{\mathbf{x}}(t) = \mathbf{A}\mathbf{x}(t) + \mathbf{B}u(t), \quad \mathbf{x}(0) = 0$$
$$y(t) = \mathbf{C}\mathbf{x}(t)$$

(2.31)

which can be represented as a transfer function. It is obtained by applying the Laplace transform to both sides of (2.31). Then, after simple transformations we have

$$[\mathbf{I}s - \mathbf{A}]\mathbf{X}(s) = \mathbf{B}U(s)$$
$$Y(s) = \mathbf{C}\mathbf{X}(s)$$

(2.32)

hence

$$\mathbf{X}(s) = [\mathbf{I}s - \mathbf{A}]^{-1}\mathbf{B}U(s)$$

(2.33)

and

$$Y(s) = \mathbf{C}[\mathbf{I}s - \mathbf{A}]^{-1}\mathbf{B}U(s).$$

(2.34)

The transfer function is represented by the following relationship, resulting from (2.34)

$$\frac{Y(s)}{U(s)} = \mathbf{C}[\mathbf{I}s - \mathbf{A}]^{-1}\mathbf{B} = \frac{\mathbf{C}\,\mathrm{adj}[\mathbf{I}s - \mathbf{A}]\mathbf{B}}{\det[\mathbf{I}s - \mathbf{A}]}$$

(2.35)

If the state equation (2.31) is given in phase-variable canonical form, then matrices $\mathbf{A}$, $\mathbf{B}$ and $\mathbf{C}$ are

$$\mathbf{A} = \begin{bmatrix} 0 & 1 & 0 & . & 0 \\ 0 & 0 & 1 & . & 0 \\ . & . & . & . & . \\ 0 & 0 & 0 & . & 1 \\ -a_0 & -a_1 & . & . & -a_{n-1} \end{bmatrix} \quad \mathbf{B} = \begin{bmatrix} 0 \\ 0 \\ . \\ . \\ 1 \end{bmatrix} \quad \mathbf{C} = \begin{bmatrix} b_0 & b_1 & . & . & b_m \end{bmatrix} \qquad (2.36)$$

and the transfer function (2.35) is given by

$$\frac{Y(s)}{U(s)} = \frac{b_m s^m + b_{m-1} s^{m-1} + \ldots + b_1 s + b_0}{s^n + a_{n-1} s^{n-1} + \ldots a_1 s + a_0}. \tag{2.37}$$

Formula (2.37) represents mutual relationships, which occur between the model expressed in transfer function form and its equivalent notation in the form of a state equation.

If the equation (2.31) is not given in phase-variable canonical form but it satisfies the conditions of controllability, i.e. the rank $\mathbf{G} = n$ where

$$\mathbf{G} = \begin{bmatrix} \mathbf{B} & \mathbf{AB} & \mathbf{A}^2\mathbf{B} & \ldots & \mathbf{A}^{n-1}\mathbf{B} \end{bmatrix} \tag{2.38}$$

then it can be brought to this form by means of a nonsingular linear transformation [30], [65], [77]:

$$\mathbf{x}(t) = \mathbf{H}\mathbf{z}(t) \tag{2.39}$$

which transforms it to the form

$$\dot{\mathbf{z}}(t) = \mathbf{A}_0 \mathbf{z}(t) + \mathbf{B}_0 u(t) \tag{2.40}$$

in which

$$\mathbf{A}_0 = \begin{bmatrix} 0 & 1 & 0 & . & 0 \\ 0 & 0 & 1 & . & 0 \\ . & & . & . & . \\ 0 & 0 & 0 & . & 1 \\ -a_0 & -a_1 & . & . & -a_{n-1} \end{bmatrix} \quad \mathbf{B}_0 = \begin{bmatrix} 0 \\ 0 \\ . \\ . \\ 1 \end{bmatrix}. \tag{2.41}$$

In order to determine the matrix $\mathbf{H}$ in (2.39), both the sides of equation (2.40) are multiplied by it, obtaining

$$\mathbf{H}\dot{\mathbf{z}}(t) = \mathbf{H}\mathbf{A}_0\mathbf{z}(t) + \mathbf{H}\mathbf{B}_0 u(t) \tag{2.42}$$

and given equation (2.39) this becomes

$$\dot{\mathbf{x}}(t) = \mathbf{H}\mathbf{A}_0\mathbf{z}(t) + \mathbf{H}\mathbf{B}_0 u(t) . \tag{2.43}$$

In turn, substituting (2.39) into (2.30) we obtain

$$\dot{\mathbf{x}}(t) = \mathbf{A}\mathbf{H}\mathbf{z}(t) + \mathbf{B}u(t) . \tag{2.44}$$

Comparing (2.43) and (2.44) yields

$$\mathbf{A}\mathbf{H} = \mathbf{H}\mathbf{A}_0 \tag{2.45}$$

and

$$\mathbf{B} = \mathbf{H}\mathbf{B}_0 \tag{2.46}$$

If matrix $\mathbf{H}$ is represented in the form of n column vectors

$$\mathbf{H} = \begin{bmatrix} \mathbf{h}_1 & \mathbf{h}_2 & \dots & \mathbf{h}_{n-1} & \mathbf{h}_n \end{bmatrix} \tag{2.47}$$

then equations (2.45) and (2.46) can be written as follows

$$\mathbf{A}\begin{bmatrix} \mathbf{h}_1 & \mathbf{h}_2 & \dots & \mathbf{h}_{n-1} & \mathbf{h}_n \end{bmatrix}$$
$$= \begin{bmatrix} \mathbf{h}_1 & \mathbf{h}_2 & \dots & \mathbf{h}_{n-1} & \mathbf{h}_n \end{bmatrix} \begin{bmatrix} 0 & 1 & 0 & . & 0 \\ 0 & 0 & 1 & . & 0 \\ . & . & . & . & . \\ 0 & 0 & 0 & . & 1 \\ -a_0 & -a_1 & . & . & -a_{n-1} \end{bmatrix} . \tag{2.48}$$

Thus

$$\begin{bmatrix} \mathbf{Ah}_1 & \mathbf{Ah}_2 & \ldots & \mathbf{Ah}_{n-1} & \mathbf{Ah}_n \end{bmatrix}$$
$$= \begin{bmatrix} -a_0\mathbf{h}_n & \mathbf{h}_1 - a_1\mathbf{h}_n & \ldots \mathbf{h}_{n-2} - a_{n-2}\mathbf{h}_n & \mathbf{h}_{n-1} - a_{n-1}\mathbf{h}_n \end{bmatrix} \tag{2.49}$$

and

$$\mathbf{B} = \begin{bmatrix} \mathbf{h}_1 & \mathbf{h}_2 & \ldots & \mathbf{h}_{n-1} & \mathbf{h}_n \end{bmatrix} \begin{bmatrix} 0 \\ 0 \\ \cdot \\ \cdot \\ 1 \end{bmatrix} = \mathbf{h}_n . \tag{2.50}$$

Comparing the respective columns in equation (2.49) with those from formula (2.50), it can be easily seen that the subsequent columns $\mathbf{h}_n$, $\mathbf{h}_{n-1}$, ... $\mathbf{h}_1$ of transformation matrix $\mathbf{H}$ determine the following recurrence formulae

$$\begin{aligned} \mathbf{h}_n &= \mathbf{B} \\ \mathbf{h}_{n-1} &= \mathbf{Ah}_n + a_{n-1}\mathbf{h}_n \\ \mathbf{h}_{n-2} &= \mathbf{Ah}_{n-1} + a_{n-2}\mathbf{h}_n \\ &\text{........} \\ \mathbf{h}_1 &= \mathbf{Ah}_2 + a_1\mathbf{h}_n . \end{aligned} \tag{2.51}$$

The solution to the state equation (2.31) for zero initial conditions is given by the formula

$$\mathbf{x}(t) = \int_0^t e^{\mathbf{A}\,(t-\tau)} \mathbf{B} u(\tau) d\tau \tag{2.52}$$

thus the response is

$$y(t) = \int_0^t \mathbf{C} e^{\mathbf{A}(t-\tau)} \mathbf{B} u(\tau)\, d\tau \tag{2.53}$$

Expression $e^{\mathbf{A}t}$ which occurs in solutions (2.52) and (2.53) represents an infinite series

$$e^{\mathbf{A}t} = \mathbf{I} + \mathbf{A}t + \frac{1}{2!}\mathbf{A}^2 t^2 + \frac{1}{3!}\mathbf{A}^3 t^3 + \ldots \tag{2.54}$$

which can be determined by various methods. In the examples we present three of these which make use of the Cayley–Hamilton theorem, inverse Laplace transform and the Sylvester formula [16], [31].

Using the Cayley–Hamilton theorem, expression $e^{\mathbf{A}t}$ can be presented by means of a finite series including powers of matrix $\mathbf{A}$, from the power zero to $(n\text{-}1)$

$$e^{\mathbf{A}t} = \sum_{k=0}^{n-1} \alpha_k \mathbf{A}^k . \tag{2.55}$$

In relationship (2.55) appear n unknown coefficients $\alpha_0, \alpha_1, \alpha_2, \ldots \alpha_{n-1}$ which satisfy the following set of equations

$$\begin{bmatrix} 1 & \lambda_1 & \lambda_1^2 & . & \lambda_1^{n-1} \\ 1 & \lambda_2 & \lambda_2^2 & . & \lambda_2^{n-1} \\ . & . & . & . & . \\ 1 & \lambda_n & \lambda_n^2 & . & \lambda_n^{n-1} \end{bmatrix} \begin{bmatrix} \alpha_0 \\ \alpha_1 \\ . \\ \alpha_{n-1} \end{bmatrix} = \begin{bmatrix} e^{\lambda_1 t} \\ e^{\lambda_2 t} \\ . \\ e^{\lambda_n t} \end{bmatrix} \tag{2.56}$$

where $\lambda_1, \lambda_2, \ldots \lambda_n$ are the eigenvalues of matrix $\mathbf{A}$. If these eigenvalues are singular then the determinant of the square matrix in (2.56) is non-zero. Hence a solution exists in the form of coefficients $\alpha_0, \alpha_1, \alpha_2, \ldots \alpha_{n-1}$ which, substituted into (2.55), gives the expression sought for $e^{\mathbf{A}t}$.

The form of $e^{\mathbf{A}t}$ can be also obtained using the Laplace transform. As there is

$$\mathscr{L}\left[e^{\mathbf{A}t}\right] = [\mathbf{I}s - \mathbf{A}]^{-1} \tag{2.57}$$

thus $e^{\mathbf{A}t}$ can be easily determined by inverse Laplace transform in equation (2.57)

$$e^{\mathbf{A}t} = \mathscr{L}^{-1}\{[\mathbf{I}s - \mathbf{A}]^{-1}\} . \tag{2.58}$$

In the Sylvester method, expression $e^{\mathbf{A}t}$ is calculated from the following formula

$$e^{\mathbf{A}t} = \sum_{k=1}^{n} \mathbf{Z}_k e^{\lambda_k t} \tag{2.59}$$

where

$$\mathbf{Z}_k = \prod_{i=1}^{n} \frac{\mathbf{A} - \mathbf{I}\lambda_i}{\lambda_k - \lambda_i} \quad i \neq k \,.$$ (2.60)

The solution of the state equation does not always require the expression $e^{\mathbf{A}t}$ to be determined. If we consider the product of matrices, denoted by $\mathbf{D}$

$$\mathbf{D} = \mathbf{P}^{-1}\mathbf{A}\mathbf{P}$$ (2.61)

where for the single eigenvalues λ_1, λ_2, ... λ_n of matrix $\mathbf{A}$, matrix $\mathbf{P}$ is in the form of

$$\mathbf{P} = \begin{bmatrix} 1 & 1 & . & 1 & 1 \\ \lambda_1 & \lambda_2 & . & \lambda_{n-1} & \lambda_n \\ \lambda_1^2 & \lambda_2^2 & . & \lambda_{n-1}^2 & \lambda_n^2 \\ . & . & . & . & . \\ \lambda_1^{n-1} & \lambda_2^{n-1} & . & \lambda_{n-1}^{n-1} & \lambda_n^{n-1} \end{bmatrix}$$ (2.62)

then relationship (2.52) can be presented in the following form which is convenient for calculations

$$\mathbf{x}(t) = \int_0^t \mathbf{P} e^{\mathbf{D}(t-\tau)} \mathbf{P}^{-1} \mathbf{B} u(\tau) d\tau \,.$$ (2.63)

Since expression $e^{\mathbf{D}(t-\tau)}$ in equation (2.63) is a diagonal matrix

$$e^{\mathbf{D}(t-\tau)} = \begin{bmatrix} e^{\lambda_1(t-\tau)} & 0 & . & 0 & 0 \\ 0 & e^{\lambda_2(t-\tau)} & . & 0 & 0 \\ . & . & . & . & . \\ 0 & 0 & . & e^{\lambda_{n-1}(t-\tau)} & 0 \\ 0 & 0 & . & 0 & e^{\lambda_n(t-\tau)} \end{bmatrix} \,.$$ (2.64)

thus solution $x(t)$ is determined without any need to calculate the expression $e^{\mathbf{A}t}$.

A similar effect of diagonalisation of matrix $\mathbf{A}$ is obtained in the case when the modal matrix $\mathbf{S}$ is used, whose columns are the eigenvectors of matrix $\mathbf{A}$

$$[\mathbf{A} - \lambda\mathbf{I}]\mathbf{x} = 0 \,.$$ (2.65)

Vector $\mathbf{x}_i$ which is the solution of equation (2.65) for the eigenvalue λ_i, is called the eigenvector. The components of vector $\mathbf{x}$ are usually normalised in such a way, so that its norm becomes unity

$$\|\mathbf{x}\| = \sqrt{\sum_{i=1}^{n}|x_i|^2} = 1 \tag{2.66}$$

If matrix $\mathbf{A}$ has single eigenvalues and matrix $\mathbf{S}$ is a modal matrix, both matrices are related

$$\mathbf{S}^{-1}\mathbf{A}\mathbf{S} = \begin{bmatrix} \lambda_1 & 0 & . & 0 & 0 \\ 0 & \lambda_2 & . & 0 & 0 \\ . & . & . & . & . \\ 0 & 0 & . & \lambda_{n-1} & 0 \\ 0 & 0 & . & 0 & \lambda_n \end{bmatrix} \tag{2.67}$$

and the solution of equation (2.31) can be written in a form being similar to that given by equation (2.63)

$$\mathbf{x}(t) = \int_{0}^{t} Se^{\mathbf{D}(t-\tau)}\mathbf{S}^{-1}\mathbf{B}u(\tau)d\tau . \tag{2.68}$$

If equation (2.1) has $(n\text{-}1)$ non-zero initial conditions then, after representing it in the form of (2.11)

$$(D^{(n)} + a_{n-1}D^{(n-1)} + ... + a_1 D + a_0)y(t) = g(t)$$
$$y(0) = y_0, \; y'(0) = y_0', \; ... \; y^{(n-1)}(0) = y_0^{(n-1)} \tag{2.69}$$

and performing the Laplace transform on (2.69) we obtain

$$[s^n Y(s) - s^{n-1}y(0) - ... - y^{(n-1)}(0)] + a_{n-1}[s^{n-1}Y(s) - s^{n-2}y(0)$$
$$ - ... - y^{(n-2)}(0)] + ... + a_0 Y(s) = G(s). \tag{2.70}$$

The equation (2.70) can be written in the form

$$[s^n + a_{n-1}s^{n-1} + ... + a_0]Y(s)$$
$$= [s^{n-1}y_0 + ... + y_0^{(n-1)}] + a_{n-1}[s^{n-2}y_0 + ... + y_0^{(n-2)}] + ... + G(s) \tag{2.71}$$

hence

$$Y(s) = \frac{[s^{n-1}y_0 + ... + y_0^{(n-1)}] + a_{n-1}[s^{n-2}y_0 + ... + y_0^{(n-2)} + ... + G(s)}{s^n + a_{n-1}s^{n-1} + ... + a_0}.$$ (2.72)

The solution to equation (2.72) depends now on determining $y(t)$ by calculating the inverse Laplace transform of $Y(s)$.

The solution of the state equation with non-zero initial conditions is given by formulae (2.73) and (2.74)

$$\mathbf{x}(t) = e^{\mathbf{A}(t-t_0)}x_0 + \int_{t_0}^{t} e^{\mathbf{A}(t-\tau)}\mathbf{B}u(\tau)\, d\tau$$ (2.73)

and

$$y(t) = \mathbf{C}e^{\mathbf{A}(t-t_0)}x_0 + \int_{t_0}^{t} \mathbf{C}e^{\mathbf{A}(t-\tau)}\mathbf{B}u(\tau)\, d\tau$$ (2.74)

2.4. Models of standards

The problem of mathematical models of standards arises in all cases pertaining to simulation investigations in the range of error determination. This applies especially in control theory and control systems as well as in measurements and recording of dynamic quantities. Models of standards can be divided into two groups. The first one constitutes standards of non-deforming signal transformation. The mathematical model of such standards is determined by means of tracing or ideal delay operations. The other group is constituted by standards of a selected objective function that is different from the non-deforming transformation. The various objective functions make it difficult to specify here a defined set of mathematical models of those standards and the mutual relations existing between them. For example, models from that group of standards can be described by totally different operations performed on the output signal, e.g. operations of averaging, filtering, differentiating, etc. Models of exemplary standards will be presented below [25], [47].

2.4.1. Models of non–deforming transformations.

Models of standards realise the tracing operation when its impulse response is in the form of

$$k_t(t) = k_0 \delta(t) \tag{2.75}$$

where k_0 is amplification of the standard and $\delta(t)$ is the Dirac delta function. A standard which satisfies equation (2.75) causes the input signal $u(t)$ to be transferred to its output after having been amplified k_0 times. This results from the convolution integral

$$y_t(t) = \int_0^t k_0 \delta(\tau) u(t - \tau)\, d\tau = k_0\, u(t). \tag{2.76}$$

From the operational transform of the standard (2.75)

$$K_t(s) = k_0 \tag{2.77}$$

results its magnitude and phase–frequency responses, shown in Fig. 2.1

$$\begin{aligned} |K_t(j\omega)| &= k_0 \\ \varphi_t(\omega) &= 0. \end{aligned} \tag{2.78}$$

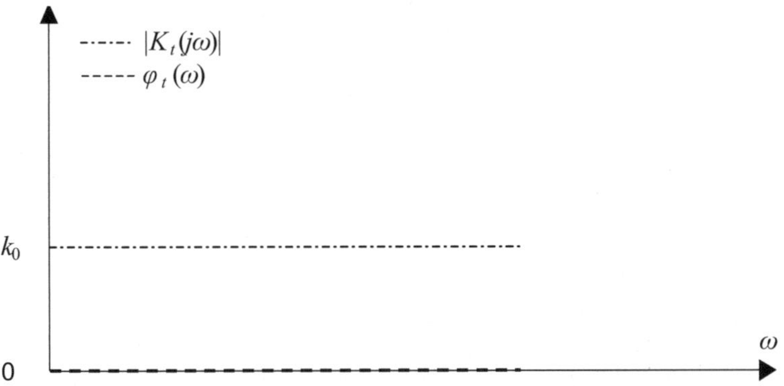

Fig. 2.1. Magnitude and phase-frequency responses of the standard $k_t(t)$ (2.75).

If the input signal $u(t)$, delayed by $\psi = \mathrm{var}$ and amplified k_0 times is transferred to the output of the standard then we say that this standard with an impulse response

$$k_s(t) = k_0 \delta(t - \psi) \qquad (2.79)$$

is suitable for evaluation of signal shape processing. By calculating the convolution integral for $k_s(t)$ as before, we now have

$$y_s(t) = \int_0^t k_0 \delta(t - \psi) u(t - \tau) d\tau = k_0 u(t - \psi) . \qquad (2.80)$$

The optimum value ψ minimises the error component resulting from the delay brought about by the physical system, however the component attributable to the signal shape processing error remains. From the operational transform (2.80) which is

$$K_s(s) = k_0 e^{-s\psi} \qquad (2.81)$$

result the magnitude and phase-frequency responses of this standard

$$|K_s(j\omega)| = k_0$$
$$\varphi_s(\omega) = -\omega\psi . \qquad (2.82)$$

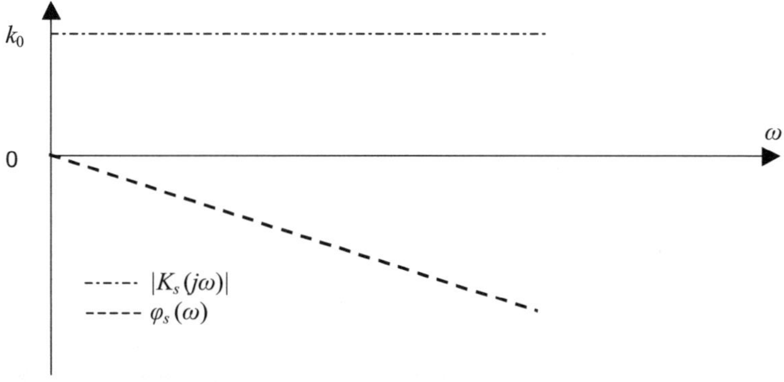

Fig. 2. 2. Magnitude and phase-frequency responses of the standard $k_s(t)$ (2.79).

2.4.2. Models of selected objective functions.

In the Examples we will consider typical standards for which the objective function is determined by averaging and filtering the signal in an ideal manner.

The model of an ideal averaging system is described by the impulse response in the form of

$$k_a(t) = \frac{k_0}{t_a}[1(t) - 1(t - t_a)] \tag{2.83}$$

where t_a is time of averaging. The response of this standard to the input signal $u(t)$ is

$$y_a(t) = \frac{k_0}{t_a}\left[\int_0^t 1(t - \tau)u(\tau)d\tau - \int_{t_a}^t 1(t + t_a - \tau)u(\tau)d\tau\right]. \tag{2.84}$$

After simple transformations which depend on writing the first integral in equation (2.84) in the form of two integrals

$$\int_0^t 1(t - \tau)u(\tau)d\tau = \int_0^{t-t_a} 1(t - \tau)u(\tau)d\tau + \int_{t-t_a}^t 1(t - \tau)u(\tau)d\tau \tag{2.85}$$

and after changing the integration limits in the second integral

$$\int_{t_a}^t 1(t + t_a - \tau)u(\tau)d\tau = \int_0^{t-t_a} 1(t - \tau)u(\tau)d\tau \tag{2.86}$$

the formula (2.84) becomes

$$y_a(t) = \frac{k_0}{t_a}\int_{t-t_a}^t u(\tau)d\tau. \tag{2.87}$$

The averaging operation, based on formula (2.87) depends on integration of the input signal within the limits $[t-t_a, t]$ and on dividing it by the averaging time t_a. The frequency responses of the averaging model result from its operational transform, which is

$$K_a(s) = \frac{k_0}{s t_a}\left(1 - e^{-s t_a}\right) \tag{2.88}$$

hence they are

$$\left|K_a(j\omega)\right| = k_0 \left|Sa\frac{\omega t_a}{2}\right|$$

$$\varphi_a(\omega) = -\frac{\omega t_a}{2} \pm n\pi.$$

(2.89)

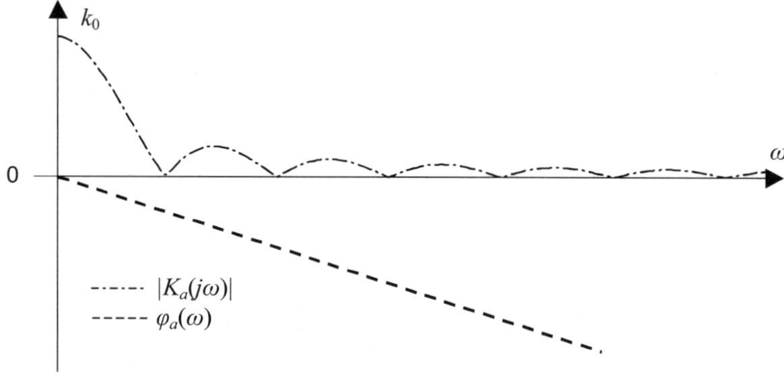

Fig. 2.3. Magnitude and phase-frequency responses of the standard $k_a(t)$ (2.83).

The mathematical model of the low–pass filter standard is described by formula

$$K_f(j\omega) = \begin{cases} k_0 e^{-j\omega t_0} & for \quad \omega < |\omega_m| \\ 0 & for \quad \omega > |\omega_m| \end{cases}$$

(2.90)

hence it has the following frequency responses

$$\left|K_f(j\omega)\right| = \begin{cases} k_0 & for \quad \omega < |\omega_m| \\ 0 & for \quad \omega > |\omega_m| \end{cases}$$

$$\varphi_f(\omega) = -\omega t_0.$$

(2.91)

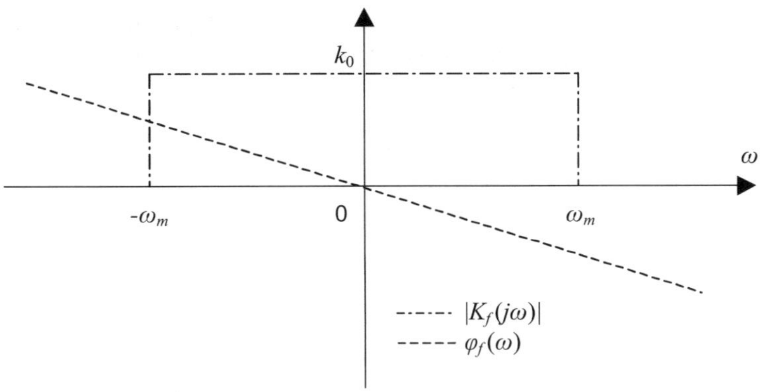

Fig. 2.4. Magnitude and phase-frequency responses of an ideal low-pass filter.

By performing an inverse Fourier transform on $K_f(j\omega)$ we obtain the impulse response $k_f(t)$ of this model

$$k_f(t) = \frac{1}{2\pi}\int\limits_{-\infty}^{\infty} K_f(j\omega)e^{j\omega t}\,d\omega = \frac{k_0}{2\pi}\int\limits_{-\omega_m}^{\omega_m} e^{j\omega(t-t_0)}\,d\omega = \frac{k_0\omega_m}{\pi}Sa[\omega_m(t-t_0)]. \quad (2.92)$$

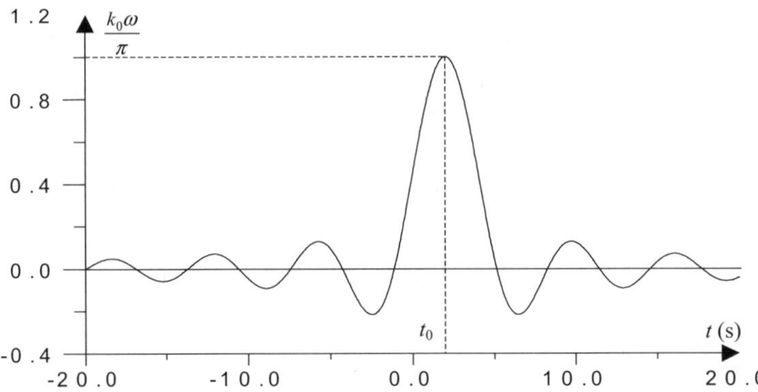

Fig. 2. 5. Impulse response of an ideal low-pass filter.

A filter described by formula (2.90) will have a maximum value of impulse response that is proportional to the limit angular frequency ω_m. If this frequency tends to infinity, the filter works across the whole band and its impulse response becomes infinitely narrow and high. It is easy to see that a filter with the model (2.90) is not physically possible, as its impulse response $k_f(t) \neq 0$ for $t < 0$.

2.5. Examples

Example 2.1

Solve the following homogeneous equation

$$y^{(5)}(t) - 3.5y^{(4)}(t) + 6y^{(3)}(t) + 4y^{(2)}(t) = 0 . \tag{2.93}$$

Solution

Characteristic equation

$$s^5 - 3.5s^4 + 6s^3 + 4s^2 = 0 \tag{2.94}$$

hence

$$s^2(s^3 - 3.5s^2 + 6s + 4) = 0 \tag{2.95}$$

has the following roots $s_1 = s_2 = 0$, $s_3 = -0.5$, $s_4 = 2 + 2j$, $s_5 = 2 - 2j$. Therefore this is the case given by formula (2.6) with respect to the root s_3, formula (2.7) with respect to the complex roots s_4 and s_5 and formula (2.8) with respect to the double root $s_1 = s_2$. The solution being sought has the form

$$y(t) = c_1 + c_2 t + c_3 e^{-0.5t} + e^{2t}(c_4 \sin 2t + c_5 \cos 2t) . \tag{2.96}$$

Example 2.2

Find the annihilator of the following function

$$g(t) = 5e^{4t} + 3\cos t . \tag{2.97}$$

Solution

Building on formula (2.20) for $\alpha = 4$, $n = 1$ we find

$$(D-4)5e^{4t} = 0.$$
(2.98)

From formula (2.23) for $\beta = 1$ we find

$$(D^2 + 1)3\cos t = 0$$
(2.99)

thus the annihilator being sought has the form

$$(D-4)(D^2 + 1)g(t) = 0.$$
(2.100)

Example 2.3

Find the general solution to the following equation

$$y''(t) - 2y'(t) + y(t) = 5e^{-2t}\sin t.$$
(2.101)

Solution

Building on formula (2.22) for $n = 1$, $\alpha = -2$, $\beta = 1$ we obtain the annihilator in the form of

$$(D^2 + 4D + 5)5e^{-2t}\sin t = 0.$$
(2.102)

Using formula (2.18) yields

$$(D^2 + 4D + 5)(D^2 - 2D + 1)y(t) = 0$$
(2.103)

which has complex roots $s_1 = -2+j$, $s_2 = -2-j$ and double roots $s_3 = s_4 = 1$. The solution of the equation which corresponds to these roots, is

$$y(t) = c_1 e^t + c_2 t e^t + c_3 e^{-2t}\cos t + c_4 e^{-2t}\sin t$$
(2.104)

where the two first expressions $c_1 e^t + c_2 t e^t$ are the solution of the homogeneous equation, while the sum $c_3 e^{-2t}\cos t + c_4 e^{-2t}\sin t$ is the particular solution.

Example 2.4

Determine the response $y(t)$ of the second-order model

$$\frac{Y(s)}{U(s)} = \frac{1}{\left(\dfrac{s}{\eta}\right)^2 + \left(2\beta\dfrac{s}{\eta}\right) + 1} \qquad (2.105)$$

for $\beta < 1$ and the unit step input $u(t) = \mathbf{1}(t)$.

Solution

For the unit step input

$$Y(s) = \frac{1}{s}\,\frac{1}{\left(\dfrac{s}{\eta}\right)^2 + \left(2\beta\dfrac{s}{\eta}\right) + 1}\,. \qquad (2.106)$$

Equating the characteristic equation to zero

$$s\left(\left(\frac{s}{\eta}\right)^2 + \left(2\beta\frac{s}{\eta}\right) + 1\right) = 0 \qquad (2.107)$$

we calculate poles: $s_1 = 0$, $s_2 = -\eta\beta - j\eta\sqrt{1-\beta^2}$, $s_3 = -\eta\beta + j\eta\sqrt{1-\beta^2}$ and the residual values which correspond to them using formula (2.26)

$$
\begin{aligned}
resY(s_1) &= 1 \\[2mm]
resY(s_2) &= \frac{-(1-\beta^2) - j\beta\sqrt{1-\beta^2}}{2(1-\beta^2)} \\[2mm]
resY(s_3) &= \frac{-(1-\beta^2) + j\beta\sqrt{1-\beta^2}}{2(1-\beta^2)}\,.
\end{aligned}
\qquad (2.108)
$$

The response $y(t)$ being sought is

$$
\begin{aligned}
y(t) = 1 &+ \frac{-(1-\beta^2) - j\beta\sqrt{1-\beta^2}}{2(1-\beta^2)}\,e^{(-\eta\beta - j\eta\sqrt{1-\beta^2})t} \\[2mm]
&+ \frac{-(1-\beta^2) + j\beta\sqrt{1-\beta^2}}{2(1-\beta^2)}\,e^{(-\eta\beta + j\eta\sqrt{1-\beta^2})t}
\end{aligned}
\qquad (2.109)
$$

and after performing the calculations it becomes

$$y(t) = 1 - \beta e^{-\eta \beta t} \frac{\sin\left(\eta\sqrt{1-\beta^2}\,t\right)}{\sqrt{1-\beta^2}} - e^{-\eta \beta t} \cos\left(\eta\sqrt{1-\beta^2}\,t\right). \tag{2.110}$$

Example 2.5

Determine the inverse Laplace transform $k(t)$ of the function $K(s)$

$$K(s) = \frac{s^2 + 2s + 1}{(s+2)^3}. \tag{2.111}$$

Solution

The characteristic equation

$$(s+2)^3 = 0 \tag{2.112}$$

has one triple pole: $s_1 = s_2 = s_3 = -2$. We calculate the residues which correspond to this pole from formula (2.28) for $r = 3$, $k = 1, 2, 3$. They are

$$i = 1, \;\; s_1 = -2, \;\; resK(s_1) = \frac{1}{(3-1)!} \frac{d^{3-1}}{ds^{3-1}} \left(s^2 + 2s + 1\right) = 1$$

$$i = 2, \;\; s_1 = -2, \;\; resK(s_2) = \frac{1}{(3-2)!} \frac{d^{3-2}}{ds^{3-2}} \left(s^2 + 2s + 1\right) = -2 \tag{2.113}$$

$$i = 3, \;\; s_1 = -2, \;\; resK(s_3) = \frac{1}{(3-3)!} \frac{d^{3-3}}{ds^{3-3}} \left(s^2 + 2s + 1\right) = 1.$$

Thus the function $k(t)$ being sought has the following form

$$k(t) = e^{-2t} - 2te^{-2t} + \frac{1}{2}t^2 e^{-2t}. \tag{2.114}$$

Example 2.6

Determine the response $y(t)$ of the following model

$$\frac{Y(s)}{U(s)} = \frac{1}{s + \dfrac{1}{T}} \tag{2.115}$$

as well as the phase shift of this response to the input function $u(t) = \cos \omega t$.

Solution

The function $\cos \omega t$ has the Laplace transform

$$\mathscr{L}(\cos \omega t) = \frac{s}{s^2 + \omega^2} \tag{2.116}$$

thus

$$Y(s) = \frac{s}{s^2 + \omega^2} \frac{1}{s + \dfrac{1}{T}} \cdot \tag{2.117}$$

The characteristic equation

$$(s^2 + \omega^2)(s + \frac{1}{T}) = 0 \tag{2.118}$$

has one real pole and two imaginary poles: $s_1 = -\dfrac{1}{T}$, $s_2 = -j\omega$, $s_3 = +j\omega$. The corresponding residues are

$$resY(s_1) = \frac{-T}{1 + \omega^2 T^2}$$

$$resY(s_2) = \frac{1}{2} \frac{T + j\omega T^2}{1 + \omega^2 T^2} \tag{2.119}$$

$$resY(s_3) = \frac{1}{2} \frac{T - j\omega T^2}{1 + \omega^2 T^2}$$

thus

$$y(t) = \frac{1}{1 + \omega^2 T^2} \left(-Te^{\frac{-t}{T}} + \frac{1}{2}\left(T + j\omega T^2\right)e^{-j\omega T} + \frac{1}{2}\left(T - j\omega T^2\right)e^{j\omega T} \right) \tag{2.120}$$

and after reductions

$$y(t) = \frac{1}{1+\omega^2 T^2}\left(-Te^{\frac{-t}{T}} + \frac{T}{1+\omega^2 T^2}(\cos\omega t + \omega T \sin\omega T) \right).$$

(2.121)

Finally we have

$$y(t) = \frac{T}{1+\omega^2 T^2}\left(-e^{\frac{-t}{T}} + \frac{\sqrt{1+\omega^2 T^2}}{1+\omega^2 T^2}\cos(\omega t - \text{arc tg}\,\omega T) \right).$$

(2.122)

The phase shift being sought is arc tg ωT .

Example 2.7

For the state equation

$$\dot{\mathbf{x}}(t) = \mathbf{A}\mathbf{x}(t) + \mathbf{B}u(t)$$

$$\mathbf{A} = \begin{bmatrix} 7 & 2 & 0 \\ 3 & 5 & -1 \\ 0 & 5 & -6 \end{bmatrix} \qquad \mathbf{B} = \begin{bmatrix} 1 \\ 1 \\ 1 \end{bmatrix} \qquad n = 3$$

(2.123)

determine matrix $\mathbf{H}$ which transforms this equation into phase-variable canonical form.

Solution

The characteristic equation of matrix $\mathbf{A}$ is

$$\lambda^3 - 6\lambda^2 - 38\lambda + 139 = 0 .$$

(2.124)

Making use of formulae (2.51) we obtain

$$\mathbf{h}_3 = \mathbf{B} = \begin{bmatrix} 1 \\ 1 \\ 1 \end{bmatrix}$$

$$\mathbf{h}_2 = \mathbf{A}\mathbf{h}_3 + a_2\mathbf{h}_3 = \begin{bmatrix} 7 & 2 & 0 \\ 3 & 5 & -1 \\ 0 & 5 & -6 \end{bmatrix}\begin{bmatrix} 1 \\ 1 \\ 1 \end{bmatrix} - 6\begin{bmatrix} 1 \\ 1 \\ 1 \end{bmatrix} = \begin{bmatrix} 3 \\ 1 \\ -1 \end{bmatrix} \tag{2.125}$$

$$\mathbf{h}_1 = \mathbf{A}\mathbf{h}_2 + a_1\mathbf{h}_3 = \begin{bmatrix} 7 & 2 & 0 \\ 3 & 5 & -1 \\ 0 & 5 & -6 \end{bmatrix}\begin{bmatrix} 3 \\ 1 \\ -7 \end{bmatrix} - 38\begin{bmatrix} 1 \\ 1 \\ 1 \end{bmatrix} = \begin{bmatrix} -15 \\ -17 \\ 9 \end{bmatrix}.$$

Matrix $\mathbf{H} = \begin{bmatrix} \mathbf{h}_1 & \mathbf{h}_2 & \mathbf{h}_3 \end{bmatrix}$ is thus

$$\mathbf{H} = \begin{bmatrix} -15 & 3 & 1 \\ -17 & 1 & 1 \\ 9 & -7 & 1 \end{bmatrix}. \tag{2.126}$$

It can be easily seen that matrix $\mathbf{A}_0$ (2.45) determined from matrix $\mathbf{H}$ takes phase-variable canonical form

$$\mathbf{A}_0 = \begin{bmatrix} -15 & 3 & 1 \\ -17 & 1 & 1 \\ 9 & -7 & 1 \end{bmatrix}^{-1}\begin{bmatrix} 7 & 2 & 0 \\ 3 & 5 & -1 \\ 0 & 5 & -6 \end{bmatrix}\begin{bmatrix} -15 & 3 & 1 \\ -17 & 1 & 1 \\ 9 & -7 & 1 \end{bmatrix} = \begin{bmatrix} 0 & 1 & 0 \\ 0 & 0 & 1 \\ -139 & 38 & 6 \end{bmatrix}. \tag{2.127}$$

Example 2.8

If matrix $\mathbf{A}$ is in the form (2.128), determine $e^{\mathbf{A}t}$ using the Cayley-Hamilton method

$$\mathbf{A} = \begin{bmatrix} 0 & 1 \\ -6 & -5 \end{bmatrix}. \tag{2.128}$$

Solution

The eigenvalues of matrix $\mathbf{A}$ result from the solution of equation

$$\begin{vmatrix} -\lambda & 1 \\ -6 & -5-\lambda \end{vmatrix} = 0 \tag{2.129}$$

that is

$$\lambda^2 + 5\lambda + 6 = 0 \tag{2.130}$$

from whence we obtain: $\lambda_1 = -3$, $\lambda_2 = -2$. Making use of equation (2.56) we determine coefficients α_0 and α_1

$$\begin{bmatrix} 1 & -3 \\ 1 & -2 \end{bmatrix} \begin{bmatrix} \alpha_0 \\ \alpha_1 \end{bmatrix} = \begin{bmatrix} e^{-3t} \\ e^{-2t} \end{bmatrix} \tag{2.131}$$

from which, after solving, we obtain

$$\begin{aligned} \alpha_0 &= 3e^{-2t} - 2e^{-3t} \\ \alpha_1 &= e^{-2t} - e^{-3t}. \end{aligned} \tag{2.132}$$

From formula (2.55) we have now

$$e^{\mathbf{A}t} = \left(3e^{-2t} - 2e^{-3t}\right) \begin{bmatrix} 1 & 0 \\ 0 & 1 \end{bmatrix} + \left(e^{-2t} - e^{-3t}\right) \begin{bmatrix} 0 & 1 \\ -6 & -5 \end{bmatrix}. \tag{2.133}$$

After calculations and reduction we obtain

$$e^{\mathbf{A}t} = \begin{bmatrix} 3e^{-2t} - 2e^{-3t} & e^{-2t} - e^{-3t} \\ -6e^{-2t} + 6e^{-3t} & -2e^{-2t} + 3e^{-3t} \end{bmatrix}. \tag{2.134}$$

Example 2.9

For matrix $\mathbf{A}$ in example 2.8, determine the expression $e^{\mathbf{A}t}$ using the Sylvester method.

Solution

According to formula (2.59), in the case in question we have

$$e^{\mathbf{A}t} = \mathbf{Z}_1 e^{\lambda_1 t} + \mathbf{Z}_2 e^{\lambda_2 t} \tag{2.135}$$

where $\mathbf{Z}_1$ and $\mathbf{Z}_2$, for eingenvalues $\lambda_1 = -3$ and $\lambda_2 = -2$ of matrix $\mathbf{A}$, are respectively

$$\mathbf{Z}_1 = \frac{\begin{bmatrix} -\lambda_2 & 1 \\ -6 & -5-\lambda_2 \end{bmatrix}}{\lambda_1 - \lambda_2} = \begin{bmatrix} -2 & -1 \\ 6 & 3 \end{bmatrix}$$

$$\mathbf{Z}_2 = \frac{\begin{bmatrix} -\lambda_1 & 1 \\ -6 & -5-\lambda_1 \end{bmatrix}}{\lambda_2 - \lambda_1} = \begin{bmatrix} 3 & 1 \\ -6 & -2 \end{bmatrix}.$$

(2.136)

Substituting matrices $\mathbf{Z}_1$ and $\mathbf{Z}_2$ into formula for $e^{\mathbf{A}t}$ we obtain

$$e^{\mathbf{A}t} = \begin{bmatrix} -2 & -1 \\ 6 & 3 \end{bmatrix} e^{-3t} + \begin{bmatrix} 3 & 1 \\ -6 & -2 \end{bmatrix} e^{-2t} \qquad (2.137)$$

hence

$$e^{\mathbf{A}t} = \begin{bmatrix} 3e^{-2t} - 2e^{-3t} & e^{-2t} - e^{-3t} \\ -6e^{-2t} + 6e^{-3t} & -2e^{-2t} + 3e^{-3t} \end{bmatrix}. \qquad (2.138)$$

The result obtained is the same as that in example 2.8.

Example 2.10

Find the solution to the state equation

$$\dot{\mathbf{x}}(t) = \mathbf{A}\mathbf{x}(t) + \mathbf{B}u(t)$$
$$y(t) = \mathbf{C}\mathbf{x}(t) \qquad (2.139)$$

if matrices $\mathbf{A}$, $\mathbf{B}$ and $\mathbf{C}$ have the form

$$\mathbf{A} = \begin{bmatrix} 0 & 1 & 0 \\ 0 & 0 & 1 \\ -6 & -11 & -6 \end{bmatrix} \quad \mathbf{B} = \begin{bmatrix} 0 \\ 0 \\ 1 \end{bmatrix} \quad \mathbf{C} = \begin{bmatrix} 4 & 5 & 6 \end{bmatrix} \text{ and } u(t) = \mathbf{1}(t). \qquad (2.140)$$

Solution

The eigenvalues of the matrix $\mathbf{A}$ result from the solution of equation

$$\begin{vmatrix} -\lambda & 1 & 0 \\ 0 & -\lambda & 1 \\ -6 & -11 & -6-\lambda \end{vmatrix} = 0 \tag{2.141}$$

that is

$$\lambda^3 + 6\lambda^2 + 11\lambda + 6 = 0 \tag{2.142}$$

thus $\lambda_1 = -1$, $\lambda_2 = -2$, $\lambda_3 = -3$. For these values the matrices $\mathbf{P}$, $\mathbf{P}^{-1}$ and $\mathbf{D} = \mathbf{P}^{-1}\mathbf{AP}$ are

$$\mathbf{P} = \begin{bmatrix} 1 & 1 & 1 \\ -1 & -2 & -3 \\ 1 & 4 & 9 \end{bmatrix}$$

$$\mathbf{P}^{-1} = \begin{bmatrix} 3 & \dfrac{5}{2} & \dfrac{1}{2} \\ -3 & -4 & -1 \\ 1 & \dfrac{3}{2} & \dfrac{1}{2} \end{bmatrix} \tag{2.143}$$

$$\mathbf{D} = \begin{bmatrix} -1 & 0 & 0 \\ 0 & -2 & 0 \\ 0 & 0 & -3 \end{bmatrix}$$

hence

$$e^{\mathbf{D}(t-\tau)} = \begin{bmatrix} e^{-1(t-\tau)} & 0 & 0 \\ 0 & e^{-2(t-\tau)} & 0 \\ 0 & 0 & e^{-3(t-\tau)} \end{bmatrix}. \tag{2.144}$$

Using formulae (2.53) and (2.63) we obtain

$$y(t) = \int_0^t [4 \quad 5 \quad 6] \begin{bmatrix} 1 & 1 & 1 \\ -1 & -2 & -3 \\ 1 & 4 & 9 \end{bmatrix}$$

$$\cdot \begin{bmatrix} e^{-(t-\tau)} & 0 & 0 \\ 0 & e^{-2(t-\tau)} & 0 \\ 0 & 0 & e^{-3(t-\tau)} \end{bmatrix} \begin{bmatrix} 3 & \dfrac{5}{2} & \dfrac{1}{2} \\ -3 & -4 & -1 \\ 1 & \dfrac{3}{2} & \dfrac{1}{2} \end{bmatrix} \begin{bmatrix} 0 \\ 0 \\ 1 \end{bmatrix} d\tau \tag{2.145}$$

$$= \frac{2}{3} - \frac{5}{2}\exp(-t) + 9\exp(-2t) - \frac{43}{6}\exp(-3t).$$

Example 2.11

Solve the state equation (2.139) from example 2.10 using the Laplace transform of expression e^{At} if matrices $\mathbf{A}$, $\mathbf{B}$ and $\mathbf{C}$ are

$$\mathbf{A} = \begin{bmatrix} 0 & 1 & 0 \\ 0 & 0 & 1 \\ -4 & -8 & -5 \end{bmatrix} \quad \mathbf{B} = \begin{bmatrix} 0 \\ 0 \\ 1 \end{bmatrix} \quad \mathbf{C} = [1 \quad 0 \quad 0] \text{ and } u(t) = \mathbf{1}(t). \tag{2.146}$$

Solution

The eigenvalues of matrix $\mathbf{A}$ result from equation

$$\begin{vmatrix} -\lambda & 1 & 0 \\ 0 & -\lambda & 1 \\ -4 & -8 & -5-\lambda \end{vmatrix} = 0 \tag{2.147}$$

that is

$$\lambda^3 + 5\lambda^2 + 8\lambda + 4 = 0 \tag{2.148}$$

from whence $\lambda_1 = -1$, $\lambda_2 = -2$, $\lambda_3 = -2$. Let us determine e^{At} using formula (2.58). We calculate

$$[\mathbf{I}s - \mathbf{A}] = \begin{bmatrix} s & -1 & 0 \\ 0 & s & -1 \\ 4 & 8 & s+5 \end{bmatrix} \tag{2.149}$$

thus

$$[\mathbf{I}s - \mathbf{A}]^{-1} = \begin{bmatrix} \dfrac{s^2+5s+8}{s^3+5s^2+8s+4} & \dfrac{s+5}{s^3+5s^2+8s+4} & \dfrac{1}{s^3+5s^2+8s+4} \\[2mm] \dfrac{-4}{s^3+5s^2+8s+4} & \dfrac{s(s+5)}{s^3+5s^2+8s+4} & \dfrac{s}{s^3+5s^2+8s+4} \\[2mm] \dfrac{-4s}{s^3+5s^2+8s+4} & \dfrac{-4(2s+1)}{s^3+5s^2+8s+4} & \dfrac{s^2}{s^3+5s^2+8s+4} \end{bmatrix}. \tag{2.150}$$

By determining the inverse Laplace transform of (2.150) we obtain

$$e^{\mathbf{A}t} = \begin{bmatrix} 4e^{-t}-2te^{-2t}-3e^{-2t} & 4e^{-t}-3te^{-2t}-4e^{-2t} & e^{-t}-te^{-2t}-e^{-2t} \\ -4e^{-t}+4te^{-2t}+4e^{-2t} & -4e^{-t}+6te^{-2t}+5e^{-2t} & -e^{-t}+2te^{-2t}+e^{-2t} \\ 4e^{-t}-8te^{-2t}-4e^{-2t} & 4e^{-t}-12te^{-2t}-4e^{-2t} & e^{-t}-4te^{-2t} \end{bmatrix}. \tag{2.151}$$

The response sought will be calculated by solving the integral

$$y(t) = \int_0^t \begin{bmatrix} 1 & 0 & 0 \end{bmatrix} \begin{bmatrix} 4e^{-(t-\tau)}-2(t-\tau)e^{-2(t-\tau)}-3e^{-2(t-\tau)} \\ -4e^{-(t-\tau)}+4(t-\tau)e^{-2(t-\tau)}+4e^{-2(t-\tau)} \\ 4e^{-(t-\tau)}-8(t-\tau)e^{-2(t-\tau)}-4e^{-2(t-\tau)} \end{bmatrix} \tag{2.152}$$

$$\begin{bmatrix} 4e^{-(t-\tau)}-3(t-\tau)e^{-2(t-\tau)}-4e^{-2(t-\tau)} & e^{-(t-\tau)}-(t-\tau)e^{-2(t-\tau)}-e^{-2(t-\tau)} \\ -4e^{-(t-\tau)}+6(t-\tau)e^{-2(t-\tau)}+5e^{-2(t-\tau)} & -e^{-(t-\tau)}+2(t-\tau)e^{-2(t-\tau)}+e^{-2(t-\tau)} \\ 4e^{-(t-\tau)}-12(t-\tau)e^{-2(t-\tau)}-4e^{-2(t-\tau)} & e^{-(t-\tau)}-4(t-\tau)e^{-2(t-\tau)} \end{bmatrix} \begin{bmatrix} 0 \\ 0 \\ 1 \end{bmatrix} d\tau.$$

Finally we obtain the result in the form of

$$y(t) = \frac{1}{4} - e^{-t} + \frac{3}{4}e^{-2t} + \frac{1}{2}te^{-2t}. \tag{2.153}$$

Example 2.12

Solve the homogeneous equation

$$y''(t) + 5y'(t) + 4y(t) = 0 \tag{2.154}$$

with the initial conditions: $y(0) = 1, \ y'(0) = 0.$

Solution

The Laplace transform of the equation has the form of

$$s^2 Y(s) - sy(0) - y'(0) + 5[sY(s) - y(0)] + 4Y(s) = 0 \tag{2.155}$$

from whence, having taken into consideration the numerical values resulting from the initial conditions and after reduction, we obtain

$$(s^2 + 5s + 4)Y(s) = s + 5 \tag{2.156}$$

that is

$$Y(s) = \frac{s+5}{s^2 + 5s + 4}. \tag{2.157}$$

The inverse transform of $Y(s)$ gives

$$y(t) = \frac{4}{3} e^{-t} - \frac{1}{3} e^{-4t}. \tag{2.158}$$

Example 2.13

Solve the non-homogeneous equation

$$y''(t) - 6y'(t) + 9y(t) = t^3 e^{3t} \tag{2.159}$$

with the initial conditions: $y(0) = 2, \ y'(0) = 6 .$

Solution

The Laplace transform of the equation has the form

$$s^2 Y(s) - sy(0) - y'(0) - 6\left[sY(s) - y(0)\right] + 9Y(s) = \frac{6}{(s-3)^4} \qquad (2.160)$$

from which, having taken into consideration the numerical values resulting from the initial conditions and after reduction, we obtain

$$Y(s) = \frac{2}{s-3} + \frac{6}{(s-3)^6} . \qquad (2.161)$$

The inverse transform of $Y(s)$ gives

$$y(t) = 2e^{3t} + \frac{1}{20}t^5 e^{3t} . \qquad (2.162)$$

Example 2.14

Solve the non-homogeneous equation

$$(D^2 + 5D + 6)y(t) = (4D + 1)u(t) \qquad (2.163)$$

with the initial conditions $y(0) = 1$, $y'(0) = 3$, $u(0) = 0$ and $u(t) = t$.

Solution

The operational equation has the form

$$(s^2 + 5s + 6)Y(s) - sy(0) - y'(0) - 5y(0) = (4s + 1)U(s) - 4u(0) \qquad (2.164)$$

that is

$$Y(s) = \frac{4s + 1}{(s^2 + 5s + 6)}U(s) + \frac{(s+5)y(0) + y'(0) - 4u(0)}{(s^2 + 5s + 6)} \qquad (2.165)$$

from which, after having taken into consideration the numerical values that result

from the initial conditions and from the transform of input function, we obtain

$$Y(s) = \frac{4s+1}{(s^2+5s+6)}\frac{1}{s^2} + \frac{s+8}{(s^2+5s+6)} \, . \tag{2.166}$$

Calculating the inverse transforms of the particular parts of $Y(s)$ we obtain

$$y(t) = \left[\frac{19}{36} + \frac{1}{6}t + \frac{11}{9}e^{-3t} - \frac{7}{4}e^{-2t}\right] + \left[-5e^{-3t} + 6e^{-2t}\right] \tag{2.167}$$

which after reduction gives

$$y(t) = \frac{19}{36} + \frac{1}{6}t + \frac{17}{4}e^{-2t} - \frac{34}{9}e^{-3t} \, . \tag{2.168}$$

3. SYSTEM PARAMETERS

The application of standard forcing functions in the form of unit step, ramp or harmonic inputs allows certain typical parameters to be determined, which characterise the chosen properties of dynamic systems. The following system parameters are considered the most important:

- overshoot,
- damping factor,
- half time,
- equivalent time delay,
- time constants,
- resonance angular frequency.

Below we will characterise these parameters and present methods used to determine them.

3.1. Overshoot

The overshoot is the distance between the first peak and the steady-state of underdamped step response. Its numerical value is determined by equating the derivative of this response to zero. Let us consider the step response $Y(s)$ of the second order system. We have then

$$Y(s) = \frac{1}{s} \frac{1}{\dfrac{s^2}{\eta^2} + 2\beta \dfrac{s}{\eta} + 1} \tag{3.1}$$

where η is the angular frequency of free oscillations, and after carrying out the inverse Laplace transform on $Y(s)$

$$y(t) = 1 - \frac{\beta}{\sqrt{1-\beta^2}} \exp(-\beta\eta\, t) \sin\sqrt{1-\beta^2}\,\eta t$$
$$- \exp(-\beta\eta\, t)\cos\sqrt{1-\beta^2}\,\eta t. \tag{3.2}$$

Oscillations in the response $y(t)$ are attenuated if $\beta < 1$. The first maximum $y_m(t)$ for which the overshoot is being determined occurs for

$$t_m = \frac{\pi}{\eta\sqrt{1-\beta^2}}.$$

(3.3)

By inserting time t_m from the later equation into formula (3.2) a relationship can be derived which describes the overshoot

$$\Delta y_m = \exp\frac{-\beta\,\pi}{\sqrt{1-\beta^2}}.$$

(3.4)

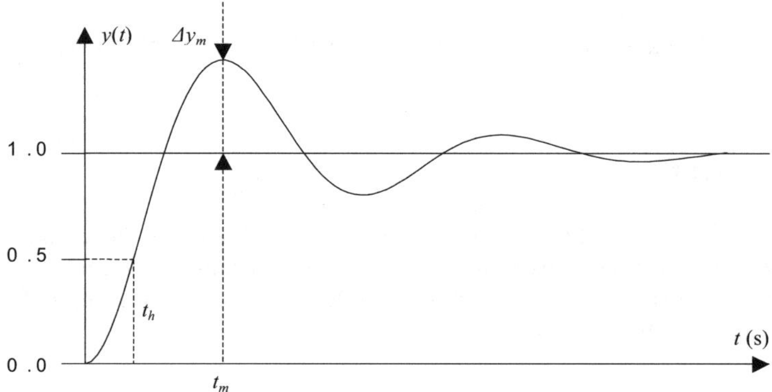

Fig. 3.1. Step response of a second-order underdamped system .

3.2. Damping factor

The damping factor β for a second order system can be determined analytically if the value of the overshoot Δy_m is given. By transforming formula (3.4) we obtain

$$\beta = \frac{\ln\Delta y_m}{\sqrt{\left(\ln\Delta y_m(t)\right)^2 + \pi^2}}.$$

(3.5)

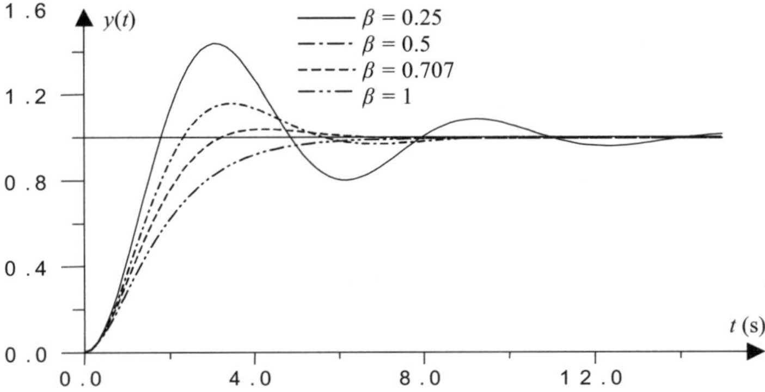

Fig. 3.2. Step responses of a second order system as a function of the damping factor β.

3.3. Half-time

Half time corresponds to a value of time $t = t_h$ for which the step response of the system first reaches half of its steady-state value - Fig 3.1. This time can be determined for a second order system using formula (3.2) and solving the equation

$$\frac{\beta}{\sqrt{1-\beta^2}}\exp(-\beta\eta\, t_h)\sin\sqrt{1-\beta^2}\,\eta t_h$$
$$+ \exp(-\beta\eta\, t_h)\cos\sqrt{1-\beta^2}\,\eta t_h = 0.5. \tag{3.6}$$

3.4. Equivalent time delay

The equivalent time delay T_{eq} corresponds to double the value of the time delay of the forced response to the ramp input

$$T_{eq} = 2\lim_{t\to\infty}\left[t - y(t)\right] \tag{3.7}$$

The force function component for a ramp input $u(t) = t$ results from the sum of the residues corresponding to the double pole $s^2 = 0$, which for a second order system

$$Y(s) = \frac{1}{s^2\left(\dfrac{s^2}{\eta^2} + 2\beta\dfrac{s}{\eta} + 1\right)} \tag{3.8}$$

are

$$\mathop{res}_{s_1=0} Y(s) = \frac{-2\beta}{\eta}$$
$$\mathop{res}_{s_2=0} Y(s) = 1.$$

(3.9)

Thus the component being sought is represented by formula

$$y(t) = t - \frac{2\beta}{\eta}$$

(3.10)

which when inserted into (3.7) gives

$$T_{eq} = \frac{4\beta}{\eta} .$$

(3.11)

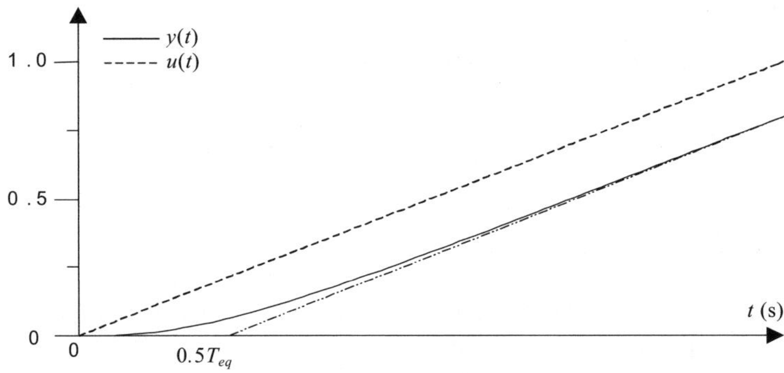

Fig. 3.3. Equivalent time delay T_{eq}.

3.5. Time constants

The time constants determine the values of the real coefficients appearing at the operator s if the denominator of the system transmittance $M(s)$ is denoted in the form of a product of first order polynomials

$$M(s) = (1 + T_1 s)(1 + T_2 s)....(1 + T_n s) .$$

(3.12)

In the case when the angular frequency of free oscillations η and the damping factor $\beta > 1$ are given, the time constants for a second order system can be determined from the relationship

$$T_{1,2} = \frac{1}{\eta}\left(\beta \pm \sqrt{\beta^2 - 1}\right) \tag{3.13}$$

and as it can easily be seen, for $\beta = 1$ the time constants are equal and they are

$$T_{1,2} = \frac{1}{\eta}. \tag{3.14}$$

3.6. Resonance angular frequency

The resonance angular frequency corresponds to a frequency at which a resonance of the magnitude-frequency response occurs. Such a response for the system (3.1) is given by the equation

$$A(\omega) = \frac{1}{\sqrt{\left(1 - \dfrac{\omega^2}{\eta^2}\right)^2 + \left(2\beta\dfrac{\omega}{\eta}\right)^2}}. \tag{3.15}$$

At resonance

$$\frac{\partial A(\omega)}{\partial \omega} = 0 \tag{3.16}$$

thus

$$\frac{2\left(1 - \dfrac{\omega^2}{\eta^2}\right)\dfrac{\omega}{\eta^2} - 4\beta^2\dfrac{\omega}{\eta^2}}{\sqrt{\left[\left(1 - \dfrac{\omega^2}{\eta^2}\right)^2 + 4\beta^2\dfrac{\omega^2}{\eta^2}\right]^3}} = 0. \tag{3.17}$$

The solution to equation (3.17) in the range of $\omega > 0$ gives the relation sought, which represents the resonance angular frequency. It is

$$\omega = \omega_r = \eta\sqrt{1 - 2\beta^2} \; .$$

(3.18)

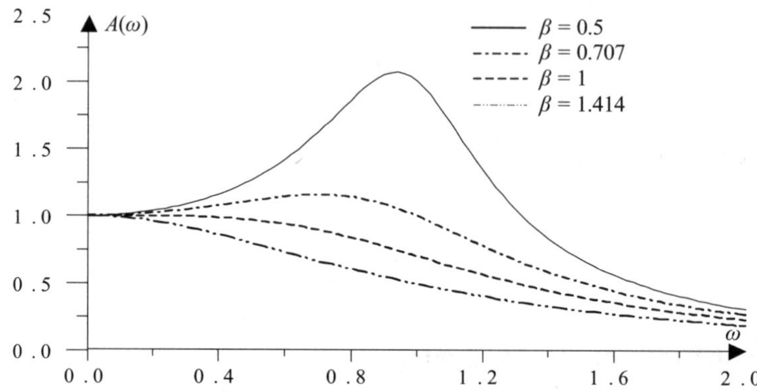

Fig. 3.4. Magnitude-frequency responses of a second order system .

4. MODEL SYNTHESIS

The synthesis of mathematical models of systems is carried out based on data obtained in the identification process. In cases where it is possible to influence input signals in a specific and deliberate way, identification is done by means of an active experiment. In many cases this is the only practical method, returning correct results in a finite time. For this purpose, an experiment plan is set up so that later on, the simplest and most effective methods of model synthesis offered by the approximation theory or the mathematical statistics, can be used. In principle, the experiment plan depends on using the optimum predetermined input signals with respect to the objective function of the model, and on a measurement of strictly defined parameters of the output signals of the system under investigation. In effect, a set of data is attained which represents the model being sought, approximated by means of a selected algorithm. The model is most often denoted by means of differential equations, state equations or transfer functions. Models presented by means of algebraic polynomials, are also important. Special attention should be given to the synthesis of the models based on the Lagrange and Chebyshev polynomials or those making use of the least square method. The possibility of transforming models expressed by means of algebraic polynomials into models in the form of transfer functions seems to be particularly attractive. The method, which makes such a transformation conceivable, has been discussed in detail in this chapter and its main advantage is the possibility of obtaining an arbitrarily high convergence of models. This chapter presents selected methods of synthesis of different time invariant models without and with the presence of disturbances. For the latter case the methods of weighted mean, smoothing functions and Kalman filter are presented.

4.1. Algebraic polynomials

4.1.1. Lagrange polynomials

Suppose the time response of the system being identified has been recorded in the form of a set of measured points (y_j, t_j). The problem of interpolation by Lagrange polynomials depends on finding such an algebraic polynomial $L_k[t, (y_j, t_j)]$ of order not higher than k, which at all interpolation points t_j would be equal to the value of y_j. Let us note that the polynomial

$$\frac{P_{k+1}(t)}{(t-t_j)\dot{P}_{k+1}(t_j)} \tag{4.1}$$

in which

$$P_{k+1}(t) = \prod_{j=0}^{k} (t - t_j) \tag{4.2}$$

and $\dot{P}_{k+1}(t_j)$ is the derivative $\dfrac{d}{dt} P_{k+1}(t)\Big|_{t=t_j}$, assumes the value of a unity

$$\frac{P_{k+1}(t)}{(t - t_j)\dot{P}_{k+1}(t_j)} = 1 \quad for \quad t = t_j \tag{4.3}$$

while at all other interpolation points it assumes the value of zero

$$\frac{P_{k+1}(t)}{(t - t_j)\dot{P}_{k+1}(t_j)} = 0 \quad for \quad t \neq t_j . \tag{4.4}$$

Hence it results that

$$L_k[t,(y_j,t_j)] = \sum_{j=0}^{k} y_j \frac{P_{k+1}(t)}{(t - t_j)\dot{P}_{k+1}(t)} \tag{4.5}$$

satisfies the conditions of the interpolation polynomial being sought. The polynomial given by formula (4.5) is called the Lagrange interpolation polynomial.

4.1.2. Chebyshev polynomials

Suppose the interpolation points t_j belong to the closed interval $[-1,1]$ and that in formula (4.5) the polynomial $P_{k+1}(t)$ is replaced by the polynomial $T_{k+1}(t)$ where

$$T_{k+1}(t) = 2tT_k(t) - T_{k-1}(t) \qquad k = 1, 2, \dots \qquad t \in [-1,1] \tag{4.6}$$

$$\begin{aligned}
T_0(t) &= \cos(0 \cdot arc\cos t) = 1 \\
T_1(t) &= \cos(1 \cdot arc\cos t) = t.
\end{aligned} \tag{4.7}$$

If, in addition, interpolation occurs in zeros of $T_{k+1}(t)$, then we say that it is performed by the Chebyshev polynomials

$$T_k[t,(y_j,t_j)] = \sum_{j=0}^{k} y_j \frac{T_{k+1}(t)}{(t-t_j)\dot{T}_{k+1}(t)} . \tag{4.8}$$

Interpolation points t_j, which determine the zeros of the $T_{k+1}(t)$ polynomials, form a triangular matrix called the experiment plan according to the zeros of the Chebyshev polynomials. For some of the initial polynomials this matrix is as follows

$$k = 1; \quad t_j = 0$$

$$k = 2; \quad t_j = -\frac{\sqrt{2}}{2}, \quad \frac{\sqrt{2}}{2}$$

$$k = 3; \quad t_j = -\frac{\sqrt{3}}{2}, \quad 0, \quad \frac{\sqrt{3}}{2}$$

$$k = 4; \quad t_j = -\frac{1}{2}\sqrt{2-\sqrt{2}}, \quad -\frac{1}{2}\sqrt{2+\sqrt{2}}, \quad \frac{1}{2}\sqrt{2-\sqrt{2}}, \quad \frac{1}{2}\sqrt{2+\sqrt{2}}$$

$$k = 5; \quad t_j = -\frac{1}{4}\sqrt{10+2\sqrt{5}}, \quad -\frac{1}{4}\sqrt{10-2\sqrt{5}},$$

$$0, \quad \frac{1}{4}\sqrt{10-2\sqrt{5}}, \quad \frac{1}{4}\sqrt{10+2\sqrt{5}}$$

$$k = 6; \quad t_j = -\frac{1}{4}\sqrt{6}-\frac{1}{4}\sqrt{2}, \quad -\frac{1}{4}\sqrt{6}+\frac{1}{4}\sqrt{2}, \quad -\frac{\sqrt{2}}{2},$$

$$\frac{\sqrt{2}}{2}, \quad \frac{1}{4}\sqrt{6}-\frac{1}{4}\sqrt{2}, \quad \frac{1}{4}\sqrt{6}+\frac{1}{4}\sqrt{2}.$$

$$\tag{4.9}$$

When the interpolation points belong to the interval $[a, b]$ then it should be transformed into the interval $[-1,1]$ using formula

$$t' = \frac{2t-a-b}{b-a} . \tag{4.10}$$

4.2. The least squares method

The approximation of the characteristic by means of Lagrange and Chebyshev polynomials leads to the synthesis of a polynomial of the order equal to the number of approximation points. Approximation by these polynomials becomes of little use if there is a large amount of measuring data (y_j, t_j). Therefore it is difficult to consider a model to be useful when it is represented by a polynomial of the order of several tens or hundreds, while the representation of measuring data by such a numerous set is not exceptional. In addition, the synthesis of a model of such a high order would require a large computational effort. It is also worth noting that measured data are always burdened with errors, which directly affect the interpolation polynomial. In such a situation it is better to construct a polynomial of relatively low order, which will pass "close" to the measuring data instead of cut "across" them. If this polynomial is such that the sum of squares of the differences between the ordinates at the measuring points $(y_j,\, t_j)$ is at minimum

$$\sum_{j=0}^{n}\left[y_j - Q(t_j)\right]^2 = \min \tag{4.11}$$

then the polynomial

$$Q(t) = \sum_{i=0}^{k} a_i f_i(t) \tag{4.12}$$

is called the least squares interpolation polynomial. A precondition for the expression (4.11) reaching minimum is that the derivatives must become equal to zero

$$\frac{\partial Q}{\partial a_0} = \frac{\partial Q}{\partial a_1} = \ldots = \frac{\partial Q}{\partial a_k} = 0 \tag{4.13}$$

which leads to the following system of equations

$$2\sum_{j=0}^{n}\left[y_j - \sum_{i=0}^{k} a_i f_i(t)\right]\left[-f_0(t_j)\right] = 0$$

$$2\sum_{j=0}^{n}\left[y_j - \sum_{i=0}^{k} a_i f_i(t)\right]\left[-f_1(t_j)\right] = 0 \tag{4.14}$$

$$\ldots\ldots\ldots\ldots$$

$$2\sum_{j=0}^{n}\left[y_j - \sum_{i=0}^{k} a_i f_i(t)\right]\left[-f_k(t_j)\right] = 0.$$

After transformation, the system of equations (4.14) denoted in normal form becomes

$$\left(\mathbf{X}^T \mathbf{X}\right)\mathbf{a} = \mathbf{X}^T \mathbf{y}. \tag{4.15}$$

In formula (4.15) matrix $\mathbf{X}$ and vectors $\mathbf{a}$ and $\mathbf{y}$ are respectively

$$
\mathbf{X} =
\begin{bmatrix}
f_0(t_0) & f_1(t_0) & f_2(t_0) & \cdot & \cdot & f_k(t_0) \\
f_0(t_1) & f_1(t_1) & f_2(t_1) & \cdot & \cdot & f_k(t_1) \\
\cdot & \cdot & \cdot & \cdot & \cdot & \cdot \\
\cdot & \cdot & \cdot & \cdot & \cdot & \cdot \\
f_0(t_n) & f_1(t_n) & f_2(t_n) & \cdot & \cdot & f_k(t_n)
\end{bmatrix}
\quad
\mathbf{a} =
\begin{bmatrix}
a_0 \\ a_1 \\ \cdot \\ a_k
\end{bmatrix}
\quad
\mathbf{y} =
\begin{bmatrix}
y_0 \\ y_1 \\ \cdot \\ y_n
\end{bmatrix}
\qquad (4.16)
$$

where vector $\mathbf{a}$, which is the solution to equation (4.15), gives the value of the sought coefficients of the polynomial $Q(t)$. Let us consider the algebraic polynomial $Q(t)$ in the form

$$
Q(t) = a_0 + a_1 t + a_2 t^2 + \ldots + a_k t^k . \qquad (4.17)
$$

For this polynomial we have

$$
f_0 = 1, \quad f_1 = t, \quad f_2 = t^2, \ldots f_k = t^k \qquad (4.18)
$$

and the matrix $\mathbf{X}$ is in the form of

$$
\mathbf{X} =
\begin{bmatrix}
1 & t_0 & t_0^2 & \cdot & \cdot & t_0^k \\
1 & t_1 & t_1^2 & \cdot & \cdot & t_1^k \\
\cdot & \cdot & \cdot & \cdot & \cdot & \cdot \\
\cdot & \cdot & \cdot & \cdot & \cdot & \cdot \\
1 & t_n & t_n^2 & \cdot & \cdot & t_n^k
\end{bmatrix} . \qquad (4.19)
$$

It is worth noting here that the success of the approximation made by the least squares method depends very much on the accuracy of all intermediate calculations. For that reason, these calculations should be carried out with maximum possible precision and the indispensable rounding up should be limited to a necessary minimum.

4.3. Cubic splines

The previous sections of this chapter dealt with the approximation of the function at close intervals by use of polynomials. Below we will present a method which can be especially useful in the approximation of such functions, for which the application of the foregoing methods would require the use of high or very high orders of polynomials e.g. those being strongly oscillatory. This method depends on splitting the given interval $[a,b]$ into a collection of subintervals $a = t_0 < t_1 < \ldots < t_n = b$ and constructing different approximating polynomials $S(t_j)$ at each subinterval. Approximation by functions of this type is called Piecewise Polynomial Approximation. The most common example of Piecewise Polynomial Approximation using cubic polynomials between each successive pair of nodes, is known as a cubic spline interpolation. In this approximation, the polynomials $S(t_j)$ are of class $C^2[a, b]$ and satisfy the following assumptions [1], [10], [54]:

$$
\begin{aligned}
a) \quad & S(t_j) = y_j \quad j = 0,1, \ldots n \\
b) \quad & S_{j+1}(t_{j+1}) = S_j(t_{j+1}) \quad j = 0,1, \ldots n - 2 \\
c) \quad & S'_{j+1}(t_{j+1}) = S'_j(t_{j+1}) \quad j = 0,1, \ldots n - 2 \\
d) \quad & S''_{j+1}(t_{j+1}) = S''_j(t_{j+1}) \quad j = 0,1, \ldots n - 2 \\
e) \quad & S''(t_0) = S''_j(t_n) = 0 \quad (\textit{free boundry condition}).
\end{aligned}
\tag{4.20}
$$

The form of the polynomial $S(t)$ at each interval $[t_{j-1}, t_j]$ can be determined after integrating twice the following equation, which results from the linearity of the second derivative,

$$
S''(t) = m_{j-1} \frac{t_j - t}{h_j} + m_j \frac{t - t_{j-1}}{h_j}
\tag{4.21}
$$

where $h_j = t_j - t_{j-1}$, $m_k = S''(t_k)$. Integrating equation (4.21) twice gives

$$
S(t) = m_{j-1} \frac{(t_j - t)^3}{6h_j} + m_j \frac{(t - t_{j-1})}{6h_j} + A_j \frac{t_j - t}{h_j} + B_j \frac{t - t_{j-1}}{h_j}.
\tag{4.22}
$$

The problem is now reduced to the determination of the integration constants A_j and B_j and the coefficients m_j. The constants A_j and B_j should be determined in such a way so that condition (4.20a) is satisfied. This can be attained by inserting $t = t_j$ and $t = t_{j-1}$ into equation (4.22). Then we have

$$
y_j = \frac{1}{6} m_j h_j^2 + B_j
\tag{4.23}
$$

and

$$
y_{j-1} = \frac{1}{6} m_{j-1} h_j^2 + A_j.
\tag{4.24}
$$

Inserting constants B_j and A_j determined from (4.23) and (4.24) into (4.22) yields

$$S(t) = m_{j-1} \frac{(t_j - t)^3}{6h_j} + m_j \frac{(t - t_{j-1})^3}{6h_j}$$
$$+ \left(y_{j-1} - \frac{m_{j-1} h_j^2}{6} \right) \frac{t_j - t}{h_j} + \left(y_j - \frac{m_j h_j^2}{6} \right) \frac{t - t_{j-1}}{h_j} \tag{4.25}$$

and

$$S'(t) = -m_{j-1} \frac{(t_j - t)^2}{2h_j} + m_j \frac{(t - t_{j-1})^2}{2h_j} + \frac{y_j - y_{j-1}}{h_j} - \frac{m_j - m_{j-1}}{6} h_j . \tag{4.26}$$

Calculating one-sided limits of the derivative from (4.26) we obtain

$$S'(t_j -) = \frac{h_j}{6} m_{j-1} + \frac{h_j}{3} m_j + \frac{y_j - y_{j-1}}{h_j}$$
$$S'(t_j +) = -\frac{h_{j+1}}{3} m_j - \frac{h_{j+1}}{6} m_{j+1} + \frac{y_{j+1} - y_j}{h_{j+1}} \tag{4.27}$$

and hence

$$S'(t_j -) = S'(t_j +) \tag{4.28}$$

thus from (4.27) results the following equation set

$$\frac{h_j}{6} m_{j-1} + \frac{h_j + h_{j+1}}{3} m_j + \frac{h_{j+1}}{6} m_{j+1} = \frac{y_{j+1} - y_j}{h_{j+1}} - \frac{y_j - y_{j-1}}{h_j} \qquad j = 1, 2, \ldots n-1 \tag{4.29}$$

which can be written in matrix form as

$$\mathbf{Am = Hy} \tag{4.30}$$

where

$$
\mathbf{A} = \begin{bmatrix}
\dfrac{h_1}{6} & \dfrac{h_1 + h_2}{3} & \dfrac{h_2}{6} & 0 & 0 & \cdots & 0 \\[2ex]
0 & \dfrac{h_2}{6} & \dfrac{h_2 + h_3}{3} & \dfrac{h_3}{6} & 0 & \cdots & 0 \\[2ex]
0 & 0 & \dfrac{h_3}{6} & \dfrac{h_3 + h_4}{3} & \dfrac{h_4}{6} & 0 & 0 \\[2ex]
\cdot & \cdot & \cdot & \cdot & \cdot & \cdot & \cdot \\[1ex]
0 & 0 & 0 & 0 & \dfrac{h_{n-1}}{6} & \dfrac{h_{n-1} + h_n}{3} & \dfrac{h_n}{6}
\end{bmatrix}
\tag{4.31}
$$

$$
\mathbf{m} = \begin{bmatrix}
m_0 \\
m_1 \\
\cdot \\
m_{n-1} \\
m_n
\end{bmatrix}
\tag{4.32}
$$

$$
\mathbf{H} = \begin{bmatrix}
\dfrac{1}{h_1} & \left(-\dfrac{1}{h_1} - \dfrac{1}{h_2}\right) & \dfrac{1}{h_2} & 0 & 0 & \cdots & 0 \\[2ex]
0 & \dfrac{1}{h_2} & \left(-\dfrac{1}{h_2} - \dfrac{1}{h_3}\right) & \dfrac{1}{h_3} & 0 & \cdots & 0 \\[2ex]
0 & 0 & \dfrac{1}{h_3} & \left(-\dfrac{1}{h_3} - \dfrac{1}{h_4}\right) & \dfrac{1}{h_4} & \cdots & 0 \\[2ex]
\cdot & \cdot & \cdot & \cdot & \cdot & \cdot & \cdot \\[1ex]
0 & 0 & 0 & 0 & \dfrac{1}{h_{n-1}} & \left(-\dfrac{1}{h_{n-1}} - \dfrac{1}{h_n}\right) & \dfrac{1}{h_n}
\end{bmatrix}
\tag{4.33}
$$

$$
\mathbf{y} = \begin{bmatrix}
y_0 \\
y_1 \\
\cdot \\
y_{n-1} \\
y_n
\end{bmatrix}.
\tag{4.34}
$$

The relationships (4.31) and (4.32) in equation (4.30) can be simplified due to the zeroing of coefficients $m_0 = m_n = 0$ resulting from condition (4.20e). In this case matrix **A** becomes a square matrix and the left-hand side of equation (4.30) assumes the form

$$\mathbf{Am} = \begin{bmatrix} \dfrac{h_1+h_2}{3} & \dfrac{h_2}{6} & 0 & 0 & ... & 0 \\[2mm] \dfrac{h_2}{6} & \dfrac{h_2+h_3}{3} & \dfrac{h_3}{6} & 0 & ... & 0 \\[2mm] 0 & \dfrac{h_3}{6} & \dfrac{h_3+h_4}{3} & \dfrac{h_4}{6} & ... & 0 \\[2mm] . & . & . & . & . & . \\[2mm] 0 & 0 & 0 & ... & \dfrac{h_{n-1}}{6} & \dfrac{h_{n-1}+h_n}{3} \end{bmatrix} \begin{bmatrix} m_1 \\ m_2 \\ . \\ m_{n-1} \end{bmatrix}. \qquad (4.35)$$

Solution of equation (4.30) with respect to $m_1, ... m_{n-1}$ allows us to determine the polynomials sought. As matrix **A** is positively determined, the values of the unknown coefficients $m_1, ... m_{n-1}$ can be explicitly found [1]. Being a polynomial of the third order, function $S(t)$ (4.25) has thus a single solution at each of the intervals $[t_{j-1}, t_j]$. The efficiency of the interpolation by means of cubic splines results from their very important property. Namely it can be shown that the set whose elements are the functions $f(t)$, having second derivatives integrable with the square at $[a, b]$, and over this interval $f(t_j) = y_j$, $j = 0,1, ... n$, the functional

$$L(f) = \int_a^b [f''(t) - S''(t)]^2 \, dt \qquad (4.36)$$

reaches minimum if $f(t)$ is approximated by means of the cubic spline $S(t)$.

4.4. Square of frequency response method

In many cases of model identification it is easy to determine the amagnitude-frequency response. It occurs particularly often for the many types of measuring systems with an electric input and output, for which a change in the frequency of the input signal does not pose any serious technical difficulties. Let us assume, for the system being identified, that such a response has been determined and the square of its response is also available. The method of synthesis of the mathematical models now depends on an approximation of the square of the response, by means of an irreducible fraction of the type

$$A^2(\omega) = \frac{b_m \omega^{2m} + b_{m-1} \omega^{2(m-1)} + ... + b_1 \omega^2 + b_0}{a_n \omega^{2n} + a_{n-1} \omega^{2(n-1)} + ... a_1 \omega^2 + a_0} \qquad m < n \quad a_n, b_m \in \Re^+. \qquad (4.37)$$

4.5. The Maclaurin series method

Below we will present a method permitting a synthesis of models, being especially useful in such cases where the systems described by these models operate in dynamic states. Thus we will require that at the beginning of the time interval, the model may precisely represent the system being identified. Let us assume that the impulse response $k_n(t)$ of such system is known and that the n-order model of this system has the form given by formula (2.24). In order to determine this model we will make use of the mutual unambiguous relations which occur between the coefficients in its numerator $b_0, b_1, \ldots b_m$ and denominator $a_0, a_1, \ldots a_{n-1}$ and the coefficients $A_{n,k}$ of the impulse response $k_n(t)$, written in the form of a Maclaurin series

$$k_n(t) = \sum_{k=0}^{\infty} \frac{1}{k!} A_{n,k}\, t^k = \sum_{k=0}^{\infty} c_{n,k} t^k \tag{4.41}$$

where the first subscript n denotes the order of the model. These relations are represented by the following matrix equation [49], [50], [64]:

$$
\begin{bmatrix}
b_m \\ b_{m-1} \\ \cdot \\ \cdot \\ b_0 \\ a_{n-1} \\ a_{n-2} \\ \cdot \\ \cdot \\ a_0
\end{bmatrix}
=
\begin{bmatrix}
\begin{pmatrix}
1 & 0 & 0 & 0 & 0 \\
0 & 1 & 0 & 0 & 0 \\
0 & 0 & 1 & 0 & 0 \\
0 & 0 & 0 & 1 & 0 \\
0 & 0 & 0 & 0 & 1 \\
0 & 0 & 0 & 0 & 0 \\
0 & 0 & 0 & 0 & 0 \\
0 & 0 & 0 & 0 & 0 \\
0 & 0 & 0 & 0 & 0 \\
0 & 0 & 0 & 0 & 0
\end{pmatrix}
\begin{pmatrix}
0 & 0 & . & 0 & 0 \\
-A_{n,0} & 0 & . & 0 & 0 \\
-A_{n,1} & -A_{n,0} & . & 0 & 0 \\
. & . & . & . & . \\
-A_{n,n-2} & -A_{n,n-3} & . & -A_{n,0} & 0 \\
-A_{n,n-1} & . & . & -A_{n,1} & -A_{n,0} \\
-A_{n,n} & . & . & -A_{n,2} & -A_{n,1} \\
. & . & . & . & . \\
. & . & . & . & . \\
-A_{n,2n-2} & . & . & -A_{n,n} & -A_{n,n-1}
\end{pmatrix}
\end{bmatrix}^{-1}
\begin{bmatrix}
A_{n,0} \\ A_{n,1} \\ A_{n,2} \\ \cdot \\ \cdot \\ \cdot \\ \cdot \\ \cdot \\ \cdot \\ A_{n,2n-1}
\end{bmatrix}
\tag{4.42}
$$

For low values of n, equation (4.42) becomes reduced to the form of

$$
\begin{bmatrix} b_0 \\ a_0 \end{bmatrix} =
\begin{bmatrix} 1 & 0 \\ 0 & -A_{10} \end{bmatrix}^{-1}
\begin{bmatrix} A_{10} \\ A_{11} \end{bmatrix}
\tag{4.43}
$$

for $n = 1$

$$
\begin{bmatrix} b_1 \\ b_0 \\ a_1 \\ a_0 \end{bmatrix} = \begin{bmatrix} 1 & 0 & 0 & 0 \\ 0 & 1 & -A_{20} & 0 \\ 0 & 0 & -A_{21} & -A_{20} \\ 0 & 0 & -A_{22} & -A_{21} \end{bmatrix}^{-1} \begin{bmatrix} A_{20} \\ A_{21} \\ A_{22} \\ A_{23} \end{bmatrix}
\tag{4.44}
$$

for $n = 2$, and

$$
\begin{bmatrix} b_2 \\ b_1 \\ b_0 \\ a_2 \\ a_1 \\ a_0 \end{bmatrix} = \begin{bmatrix} 1 & 0 & 0 & 0 & 0 & 0 \\ 0 & 1 & 0 & -A_{30} & 0 & 0 \\ 0 & 0 & 1 & -A_{31} & -A_{30} & 0 \\ 0 & 0 & 0 & -A_{32} & -A_{31} & -A_{30} \\ 0 & 0 & 0 & -A_{33} & -A_{32} & -A_{31} \\ 0 & 0 & 0 & -A_{34} & -A_{33} & -A_{32} \end{bmatrix}^{-1} \begin{bmatrix} A_{30} \\ A_{31} \\ A_{32} \\ A_{33} \\ A_{34} \\ A_{35} \end{bmatrix}
\tag{4.45}
$$

for $n = 3$, and so on.

It is worth noting here that if model (2.24) is an irreducible rational function, then the square matrices in the formulae (4.43), (4.44), ...etc. are non singular. The coefficients $A_{n,0}, A_{n,1}, \ldots A_{n,2n-1}$, which are elements of square matrices (4.42)-(4.45) represent the initial conditions of the function $k_n(t)$ at zero. Thus we have

$$
A_{n,0} = k_n(t)\big|_{t=0}, \quad A_{n,1} = \frac{dk_n(t)}{dt}\bigg|_{t=0},
$$

$$
\frac{A_{n,2}}{2!} = \frac{d^2 k_n(t)}{dt^2}\bigg|_{t=0}, \quad \cdots \quad \frac{A_{n,k}}{k!} = \frac{d^k k_n(t)}{dt^k}\bigg|_{t=0}.
\tag{4.46}
$$

From relationship (4.42) we know that, in order to determine all parameters $a_{n,0}, a_{n,1}, \ldots a_{n,n-1}$ and $b_{n,0}, b_{n,1}, \ldots b_{n,m}$ of the n-th order model (2.24), it is necessary to know the values of the $2n$ initial coefficients $A_{n,0}, A_{n,1}, \ldots A_{n,2n-1}$. It is easy to obtain these coefficients by determining the consecutive derivatives according to formula (4.46). They can also be easily determined using the regression method. This is done by minimising a selected functional, which is determined on the difference $k_n(t)$ and the Maclaurin series limited to the $2n$ initial expressions. Let us consider the square functional in the form of

$$
I_2 = \int_0^T \left(k_n(t) - \sum_{k=0}^{2n-1} \frac{1}{k!} A_{n,k} t^k \right)^2 dt .
\tag{4.47}
$$

The minimisation of (4.47) is done by calculating derivatives

$$\frac{\partial I_2}{\partial A_{n,k}} = 0 \qquad k = 0, 1, \ldots 2n - 1 \tag{4.48}$$

which leads to a system of $2n$ equations

$$\frac{2}{k!} \int_0^T \left(k_n(t) - \sum_{k=0}^{2n-1} \frac{1}{k!} A_{n,k} t^k \right) t^k \, dt = 0 \qquad k = 0, 1, \ldots 2n - 1 \tag{4.49}$$

where T is the time in which the minimisation is performed. Solution of equations (4.49) allows the sought coefficients $A_{n,0}, A_{n,1}, \ldots A_{n,2n-1}$ of the power series (4.41) to be determined. The relationship which permits the coefficients $A_{n,0}, A_{n,1}, \ldots A_{n,2n-1}$ of the impulse response of the system to be calculated on the basis of the knowledge of its parameters $a_0, a_1, \ldots a_{n-1}$, $b_0, b_1, \ldots b_m$ is shown in equation

$$
\begin{bmatrix}
A_{n,0} \\
A_{n,1} \\
A_{n,2} \\
\cdot \\
\cdot \\
\cdot \\
\cdot \\
A_{n,2n-1}
\end{bmatrix}
=
\begin{bmatrix}
1 & 0 & 0 & 0 & \cdot & 0 & 0 & 0 \\
a_{n-1} & 1 & 0 & 0 & \cdot & 0 & 0 & 0 \\
a_{n-2} & a_{n-1} & 1 & 0 & \cdot & 0 & 0 & 0 \\
\cdot & \cdot & \cdot & \cdot & \cdot & \cdot & \cdot & \cdot \\
a_0 & a_1 \cdot & a_2 & a_3 & \cdot & 0 & 0 & 0 \\
0 & a_0 & a_1 & a_2 & \cdot & 1 & 0 & 0 \\
0 & 0 & a_0 & a_1 & \cdot & a_{n-1} & 1 & 0 \\
0 & 0 & 0 & a_0 & \cdot & a_{n-2} & a_{n-1} & 1
\end{bmatrix}^{-1}
\begin{bmatrix}
b_m \\
b_{m-1} \\
b_{m-2} \\
\cdot \\
b_0 \\
0 \\
\cdot \\
0
\end{bmatrix} \tag{4.50}
$$

The subsequent coefficients of the series $A_{n,2n}, A_{n,2n+1}, A_{n,2n+2}, \ldots$ of the first column of equation (4.50) are expressed by the coefficients preceding them $A_{n,0}, A_{n,1}, \ldots A_{n,2n-1}$. The following relations are dependent on the assigned order n:

$$
\begin{bmatrix}
A_{12} \\
A_{13} \\
A_{14} \\
\cdot \\
\cdot
\end{bmatrix}
=
\begin{bmatrix}
-A_{11} \\
-A_{12} \\
-A_{13} \\
\cdot \\
\cdot
\end{bmatrix} a_0 \tag{4.51}
$$

for $n = 1$

$$\begin{bmatrix} A_{24} \\ A_{25} \\ A_{26} \\ \cdot \\ \cdot \end{bmatrix} = \begin{bmatrix} -A_{23} & -A_{22} \\ -A_{24} & -A_{23} \\ -A_{25} & -A_{24} \\ \cdot & \cdot \end{bmatrix} \begin{bmatrix} a_1 \\ a_0 \end{bmatrix} \qquad (4.52)$$

for $n = 2$

$$\begin{bmatrix} A_{36} \\ A_{37} \\ A_{38} \\ \cdot \\ \cdot \end{bmatrix} = \begin{bmatrix} -A_{35} & -A_{34} & -A_{33} \\ -A_{36} & -A_{35} & -A_{34} \\ -A_{37} & -A_{36} & -A_{35} \\ \cdot & \cdot & \cdot \end{bmatrix} \begin{bmatrix} a_2 \\ a_1 \\ a_0 \end{bmatrix} \qquad (4.53)$$

for $n = 3$ and so on, from which it results that all information describing function $k_n(t)$ is contained in the finite number $2n$ of the power series expansion initial coefficients.

4.6. Multi-inertial models

A large class of models, especially for various heat systems, are multi-inertial models. We will present now a method for model synthesis for such systems, based on the first part of their step responses and on the solution to equation (4.50) with respect to the coefficients $A_{n,0}, A_{n,1}, \dots A_{n,2n-1}$. It is assumed that the system is linear and the model will be of the form (2.24). It is also assumed that there is given a set of discrete values of the step response of this system, in the time interval from zero to reaching the flex point. From formula (4.50) it can be easily seen that the coefficients $A_{n,0}, A_{n,1}, \dots A_{n,2n-1}$ represent the following relationship

$$A_{n,0} = b_m$$

$$A_{n,1} = -b_m a_{n-1} + b_{m-1}$$

$$A_{n,2} = b_m (a_{n-1})^2 - b_m a_{n-2} - b_{m-1} a_{n-1} + b_{m-2}$$

$$A_{n,3} = -b_m (a_{n-1})^3 + 2 b_m a_{n-2} a_{n-1} - b_m a_{n-3}$$
$$+ b_{m-1} (a_{n-1})^2 - b_{m-1} a_{n-2} - b_{m-2} a_{n-1} + b_{m-3}$$

$$A_{n,4} = -b_m (a_{n-1})^4 - 3 b_m a_{n-2} (a_{n-1})^2 + 2 b_m a_{n-3} a_{n-1}$$
$$+ b_m (a_{n-2})^2 - b_m a_{n-4} - b_{m-1} (a_{n-1})^3 + 2 b_{m-1} a_{n-2} a_{n-1}$$
$$- b_{m-1} a_{n-3} + b_{m-2} (a_{n-1})^2 - b_{m-2} a_{n-2} - b_{m-3} a_{n-1} + b_{m-4}$$

$$\tag{4.54}$$

....

etc.

The impulse response of model (2.24) represented as a power series has the form

$$k(t) = A_{n,0} \frac{t^0}{0!} + A_{n,1} \frac{t^1}{1!} + A_{n,2} \frac{t^2}{2!} + A_{n,3} \frac{t^3}{3!} + \ldots \ . \tag{4.55}$$

We then obtain the step response $h(t)$ by integrating $k(t)$

$$h(t) = A_{n,0} \frac{t^1}{1!} + A_{n,1} \frac{t^2}{2!} + A_{n,2} \frac{t^3}{3!} + A_{n,3} \frac{t^4}{4!} + \ldots \ . \tag{4.56}$$

The responses $k(t)$ and $h(t)$ can be expressed by the coefficients of the numerator $b_0, b_1, \ldots b_m$ and of the denominator $a_0, a_1, \ldots a_{n-1}$ of model (2.24) after taking into account relationship (4.54) in formulae (4.55) and (4.56). It is then easy to see that if the difference between the orders of the numerator and denominator is r, then the step response $h(t)$ can be represented by the following relationship

$$h(t) = A_{n,0} \frac{t^r}{r!} + A_{n,1} \frac{t^{r+1}}{(r+1)!} + A_{n,2} \frac{t^{r+2}}{(r+2)!} + A_{n,3} \frac{t^{r+3}}{(r+3)!} + \ldots \ . \tag{4.57}$$

The difference $r = n - m$ is called the reduced order of the model. The approximate relationship which occurs between the relation of two discrete values of the response $h(t)$ in $t_1 = \Delta$ and $t_2 = 2\Delta$ is as follows [74]:

$$\frac{h(2\Delta)}{h(\Delta)} \approx 2^r \Phi \tag{4.58}$$

where Φ is determined by formula

$$\Phi = \frac{1 - \left[1 - \dfrac{b_{m-1}}{b_m}\left(\dfrac{\overline{T}}{n}\right)\right]\cdot\dfrac{2n}{r+1}\left(\dfrac{\Delta}{\overline{\overline{T}}}\right)}{1 - \left[1 - \dfrac{b_{m-1}}{b_m}\left(\dfrac{\overline{T}}{n}\right)\right]\cdot\dfrac{n}{r+1}\left(\dfrac{\Delta}{\overline{\overline{T}}}\right)}\cdots$$

$$\cdots\frac{+\left[1 - \dfrac{b_{m-1}}{b_m}\left(\dfrac{\overline{T}}{n}\right) + \left(\dfrac{b_{m-2}}{b_m} - a_{n-2}\right)\left(\dfrac{\overline{T}}{n}\right)^2\right]\cdot\dfrac{4n^2}{(r+1)(r+2)}\left(\dfrac{\Delta}{\overline{\overline{T}}}\right)^2 + \ldots}{+\left[1 - \dfrac{b_{m-1}}{b_m}\left(\dfrac{\overline{T}}{n}\right) + \left(\dfrac{b_{m-2}}{b_m} - a_{n-2}\right)\left(\dfrac{\overline{T}}{n}\right)^2\right]\cdot\dfrac{n^2}{(r+1)(r+2)}\left(\dfrac{\Delta}{\overline{\overline{T}}}\right)^2 + \ldots} \tag{4.59}$$

and $\overline{T}$ is the mean time constant which satisfies the relation

$$\overline{T} = \frac{n}{\displaystyle\sum_{i=1}^{n}\frac{1}{T_i}}\,. \tag{4.60}$$

Thus for a model with stable poles there is

$$\frac{1}{\overline{\overline{T}}} = \frac{-(\delta_1 + \delta_2 + \ldots + \delta_n)}{n} \tag{4.61}$$

where $\delta_j = \mathrm{Re}[s_j]$ and s_j are the poles of the model. On the basis of (4.60) and using the properties of function Φ it is easy to determine the reduced r order of the model. Finding the logarithm of (4.58) we obtain

$$\ln\frac{h(2\Delta)}{h(\Delta)} \approx r\ln 2 + \ln \Phi\,. \tag{4.62}$$

However, as the function $\Phi \approx 1$ for relatively small sampling periods $\dfrac{\Delta}{\overline{\overline{T}}}$, equation (4.62) is reduced to

$$\frac{\ln \dfrac{h(2\Delta)}{h(\Delta)}}{\ln 2} = r_0 \approx r \tag{4.63}$$

from whence after rounding up r_0 to an integer we can quite accurately evaluate the unknown value of r.

A special case of the model (2.24) is the multi-inertial model of Strejc

$$K_s(s) = \frac{k}{(1+s\overline{T})^n} \qquad \overline{T} = T_i \quad i = 1, 2, \dots n. \tag{4.64}$$

For model (4.64) the order of the numerator is $m = 0$. Its denominator order $n = r$ results from equation (4.63), and is determined on the basis of sample measurements satisfying the relationship

$$n = r = Ent\{r_0\}. \tag{4.65}$$

Having at our disposal the model order n, the mean time constant $\overline{T}$ and amplification k should be determined in order to identify its remaining parameters. The relationships, which allow those parameters to be computed, are as follows [74]

$$\overline{T} \approx \frac{n}{\dfrac{n+1}{\Delta}\left[2^n \dfrac{h(\Delta)}{h(2\Delta)} - 1\right]} \tag{4.66}$$

$$k \approx \frac{h^2(\Delta)}{h(2\Delta)}\left(\frac{2}{\Delta}\right)^n n!\overline{T}^n. \tag{4.67}$$

The synthesis of the Strejc model can easily be done on the basis of measurements of the step response value $h(\Delta)$ and $h(2\Delta)$ and using equations (4.63) - (4.67).

The multi-inertial model without differentiation, with various time constants T_i for $i = 1, 2, \dots n$ is of a more general character

$$K(s) = \frac{k}{\prod\limits_{i=1}^{n}(1+sT_i)}. \tag{4.68}$$

For this model, the amplification coefficient k is being determined on the basis of (4.67) and the time constants T_i, $i = 1, 2, \ldots n$ are being determined for the highest time constant $T_i = T_{max}$ and the scatter coefficient λ

$$T_i = \lambda^{i-1} T_{max} \qquad i = 1, 2, \ldots n \tag{4.69}$$

where

$$\lambda = \sqrt[n-1]{\frac{T_{min}}{T_{max}}} \qquad 0 < \lambda \leq 1, \tag{4.70}$$

$$T_{max} = \max\left[-\frac{1}{\delta_1}, \ldots -\frac{1}{\delta_n}\right], \tag{4.71}$$

$$T_{min} = \min\left[-\frac{1}{\delta_1}, \ldots -\frac{1}{\delta_n}\right]. \tag{4.72}$$

The value T_{max} in (4.69) is estimated on the basis of measurements of the response in $h(t_1)$ and $h(t_1 + \Delta t)$ from the formula

$$T_{max} \approx \frac{\Delta t}{\ln \dfrac{k - h(t_1)}{k - h(t_1 + \Delta t)}}. \tag{4.73}$$

Having the model order n, the mean time constant $\overline{T}$ and the maximum time constant T_{max} at our disposal we can evaluate the scatter coefficient, thus solving the following equation

$$\lambda^{n-1}\left(1 - n\frac{T_{max}}{\overline{T}}\right) + \lambda^{n-2} + \ldots + \lambda + 1 = 0 \tag{4.74}$$

which results from (4.60) and (4.69).

4.7. Weighted means method

In the case when the impulse response $k(t)$ is interfered with by the noise $z(t)$, determination of coefficients A_k of series (4.41) as well as determination of the coefficients $a_0 - a_{n-1}$ and $b_0 - b_m$ of model (2.24) by means of formula (4.42), presents a series of difficulties related to the differentiation of the noise. In order to avoid these difficulties, the value of the derivatives of the noisy signals can be determined with good approximation using the weighted mean [33], [60]:

$$\bar{k}(t) = \frac{\int\limits_{t-\delta}^{t+\delta} k(\tau)\, g(\tau - t)\, d\tau}{\int\limits_{t-\delta}^{t+\delta} g(\tau - t)\, d\tau} \tag{4.75}$$

where $\bar{k}(t)$ is the weighted mean, $g(\tau - t)$ is the weight function, 2δ is the width of the intervals of averaging. The properties of averaging depend on the width of the interval 2δ and on the form of the function $g(\tau - t)$. They should be chosen in such a way so that they would assume zero values at the averaging intervals ends $(t - \delta)$, $(t + \delta)$ and a maximum value in the middle of it. If we also assume for $g(\tau - t)$ a function whose subsequent derivatives with regard to τ will become equal to zero at the ends of the interval $(t - \delta)$, $(t + \delta)$, then the k-th derivative of the weighted mean can be determined from the following formula

$$\bar{k}^{(k)}(t) = \frac{(-1)^k \int\limits_{t-\delta}^{t+\delta} k(\tau)\, g^{(k)}(\tau - t)\, d\tau}{\int\limits_{t-\delta}^{t+\delta} g(\tau - t)\, d\tau} \tag{4.76}$$

from which it results that the differentiation operation is transferred on the function of weight. Formula (4.76) can be easily determined when calculating the integral of the first derivative of function $\bar{k}(t)$. We then have

$$\bar{k}'(t) = \frac{\int\limits_{t-\delta}^{t+\delta} k'(\tau)g(\tau-t)d\tau}{\int\limits_{t-\delta}^{t+\delta} g(\tau-t)d\tau} = \frac{k(\tau)g(\tau-t)\big|_{t-\delta}^{t+\delta}}{\int\limits_{t-\delta}^{t+\delta} g(\tau-t)d\tau} - \frac{\int\limits_{t-\delta}^{t+\delta} k(\tau)g'(\tau-t)d\tau}{\int\limits_{t-\delta}^{t+\delta} g(\tau-t)d\tau} \qquad (4.77)$$

however, because of the assumption

$$g(\tau-t)\big|_{t-\delta}^{t+\delta} = 0 \qquad (4.78)$$

thus

$$\bar{k}'(t) = \frac{(-1)\int\limits_{t-\delta}^{t+\delta} k(\tau)g'(\tau-t)d\tau}{\int\limits_{t-\delta}^{t+\delta} g(\tau-t)d\tau}. \qquad (4.79)$$

Repeating identical calculations for the second and further derivatives we obtain formula (4.76). Exemplary weight functions $g(\tau-t)$, becoming zero at the ends of averaging intervals and having there $2n$ derivatives equal to zero, are represented by Nuttall windows

$$g(\tau-t) = \cos^{P}\left[\frac{\pi}{2\delta}(\tau-t)\right] \qquad p = 1,2,3, \dots \qquad (4.80)$$

or windows of the type

$$g(\tau-t) = \left[1-\left|\frac{\tau-t}{\delta}\right|\right]^{p} \qquad p = 1,2,3, \dots \, . \qquad (4.81)$$

In order to simplify calculations, it is convenient to normalise the weighted mean, bringing its denominator to 1. If we denote the denominator of the fraction in formula (4.76) by d

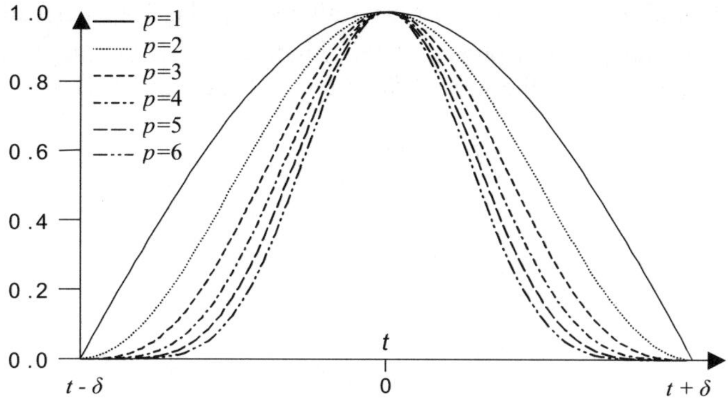

Fig. 4.1. Nuttall windows $g(\tau - t)$ (4.80) for $p = 1 \div 6$.

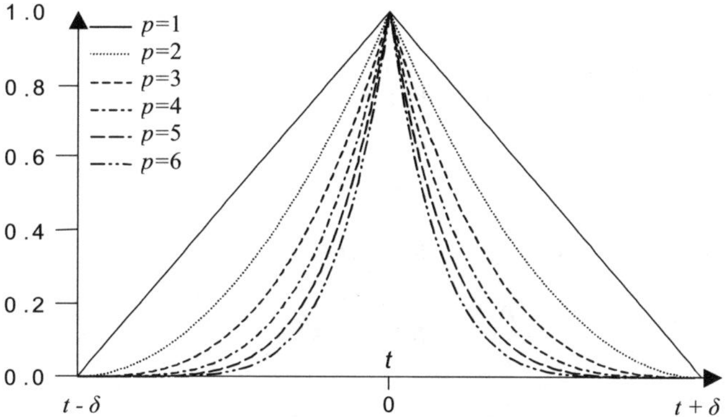

Fig. 4.2. Windows $g(\tau - t)$ (4.81) for $p = 1 \div 6$.

$$d = \int_{t-\delta}^{t+\delta} g(\tau - t)d\tau \tag{4.82}$$

then the normalised weighted mean is given as

$$\bar{k}^{(k)}(t) = (-1)^k d^{-1} \int_{t-\delta}^{t+\delta} [k(\tau) g^{(k)}(\tau - t)]d\tau \tag{4.83}$$

Substituting the ordinates from the response $A^2(\omega)$ for $n+m+2$ values of ω, into expression (4.37), we obtain a system of $n+m+2$ linear equations with respect to the unknown coefficients $a_0, \ldots a_n, b_0, \ldots b_m$ in the form

$$A^2(\omega_1)\sum_0^n a_n\omega_1^{2n} - \sum_0^m b_m\omega_1^{2m} = 0$$

$$A^2(\omega_2)\sum_0^n a_n\omega_2^{2n} - \sum_0^m b_m\omega_2^{2m} = 0 \qquad (4.38)$$

$$\ldots\ldots\ldots\ldots$$

$$A^2(\omega_{n+m+2})\sum_0^n a_n\omega_{n+m+2}^{2n} - \sum_0^m b_m\omega_{n+m+2}^{2m} = 0.$$

Solution of equations (4.38) gives the values of $a_0, \ldots a_n, b_0, \ldots b_m$ sought. As the square of response $A^2(\omega)$ is an even function of the argument, it can be written in the following form

$$A^2(\omega) = K(j\omega)K^*(j\omega) = \left(\frac{\sqrt{b_m}}{\sqrt{a_n}}\frac{(\omega-\lambda_1)\ldots(\omega-\lambda_m)}{(\omega-s_1)\ldots(\omega-s_n)}\right)$$
$$\cdot\left(\frac{\sqrt{b_m}}{\sqrt{a_n}}\frac{(\omega-\lambda_1^*)\ldots(\omega-\lambda_m^*)}{(\omega-s_1^*)\ldots(\omega-s_n^*)}\right) \qquad (4.39)$$

where λ_i is the i-th zero of the numerator with a positive imaginary part, s_k is the k-th pole of the denominator with a positive imaginary part and $K^*(j\omega), \lambda^*, s^*$ are conjugates to $K(\omega)$, λ, and s respectively. If we assume that the system being modelled is minimum-phase, then its poles and zeros are in the upper half-plane of ω, hence the multiplicand of equation (4.39) corresponds to the function $K(j\omega)$

$$K(j\omega) = \left(\frac{\sqrt{b_m}}{\sqrt{a_n}}\frac{(\omega-\lambda_1)\ldots(\omega-\lambda_m)}{(\omega-s_1)\ldots(\omega-s_n)}\right). \qquad (4.40)$$

The transfer function $K(s)$ is obtained from formula (4.40) by transferring all poles and zeros from the upper half-plane of ω into the left half-plane of s. We obtain it by multiplying the numerator of fraction (4.40) by j^m, its denominator by j^n and inserting $s = j\omega$.

where the condition $p \geq k-1$ must be satisfied by the k-th derivative of the weighted mean and the p-th power of the window. It can be easily shown that some initial values of d for the weight function (4.80), are as follows

$$
\begin{aligned}
d &= \frac{4\delta}{\pi} && for \quad p=1 \\[2mm]
d &= \delta && for \quad p=2 \\[2mm]
d &= \frac{8\delta}{\pi} && for \quad p=3 \\[2mm]
d &= \frac{3\delta}{4} && for \quad p=4 \\[2mm]
d &= \frac{32\delta}{15\pi} && for \quad p=5 \\[2mm]
d &= \frac{5\delta}{8} && for \quad p=6
\end{aligned}
\tag{4.84}
$$

while for the weight function (4.81) they are

$$
\begin{aligned}
d &= \delta && for \quad p=1 \\[2mm]
d &= \frac{2\delta}{3} && for \quad p=2 \\[2mm]
d &= \frac{\delta}{2} && for \quad p=3 \\[2mm]
d &= \frac{2\delta}{5} && for \quad p=4 \\[2mm]
d &= \frac{\delta}{3} && for \quad p=5 \\[2mm]
d &= \frac{2\delta}{7} && for \quad p=6.
\end{aligned}
\tag{4.85}
$$

The values of the coefficients A_k (4.41) based on the weighted mean we obtain by dividing the result of relationship (4.83) by k!

$$
A_k = \frac{\overline{k}^{(k)}(t)}{k!} \qquad k = 0,1,2,\ \dots\ .
\tag{4.86}
$$

When $k(t)$ is given in analytical form and hence $k(-t)$ is also known, numerical computation of the integral in formula (4.83) can be done within the limits

$[-\delta,+\delta]$, assuming $t = 0$. However, if $k(t) \equiv 0$ for $t < 0$, the time t at the lower limits of integration should be shifted by $t \geq \delta$, thus fixing this limit at zero or at positive values. Of course such a shift of the limit of integration introduces some errors in computing the values of coefficients, which instead of $t = 0$, will be computed for t shifted at least at the window width δ.

If interferences $z(t)$ occur, i.e.

$$k_z(t) = k(t) + z(t) \tag{4.87}$$

the relationship (4.83) is a sum of two integrals

$$\bar{k}_z^{(k)}(-1)^k d^{-1} \int_{t-\delta}^{t+\delta} [k(\tau)g^{(k)}(\tau-t)]d\tau + (-1)^k d^{-1} \int_{t-\delta}^{t+\delta} [z(\tau)g^{(k)}(\tau-t)]d\tau \tag{4.88}$$

in which the differentiation of $z(\tau)$ was transferred on the function of weight. If we assume that the interfering signal $z(\tau)$ is a random signal varying its value and sign quickly with respect to $g^{(k)}(\tau-t)$, then from evaluation of the second integral in formula (4.88) we obtain

$$d^{-1} \int_{t-\delta}^{t+\delta} [z(\tau)g^{(k)}(\tau-t)]d\tau \leq \sup_{t-\delta \leq \tau \leq t+\delta} [g^{(k)}(\tau-t)]d^{-1} \int_{t-\delta}^{t+\delta} [z(\tau)]d\tau . \tag{4.89}$$

From the assumption adopted with respect to the signal $z(\tau)$ and for a sufficiently high window width δ there is

$$\int_{t-\delta}^{t+\delta} [z(\tau)]d\tau \cong 0 \tag{4.90}$$

which means attenuation of interferences. The weighted mean of the interference signal is thus represented by the approximated relationship

$$\bar{k}_z^{(k)} \cong (-1)^k d^{-1} \int_{t-\delta}^{t+\delta} [k(\tau)g^{(k)}(\tau-t)]d\tau \tag{4.91}$$

while the coefficients A_k of the series (4.41) are given by the formula

$$A_k \cong \frac{\overline{k}_z^{(k)}(t)}{k!} \qquad k = 0,1,2, \ldots \ .$$

(4.92)

4.8. Smoothing functions

Now we will consider a similar case where the disturbed system response is determined at the interval $[a,b]$. Let us assume that the values of the ordinates $\widetilde{y}_j$ of the function being approximated $\widetilde{f}(t_j)$ are burdened with errors, and that the ordinates $\widetilde{y}_j = \widetilde{f}(t_j)$ correspond to the abscises t_j, $t_j \in [a,b]$. We will look for an approximating function with ordinates $y_j = f(t_j)$, whose shape in the vicinity of t_j is smoother than that of the approximated function and the error caused by interference is at a minimum. Such functions are called smoothing functions. It is required that those functions minimise the following functional [56]:

$$L(f) = \int_a^b [f''(t)]^2 \, dt + \sum_{k=0}^n p_j [f(t_j) - \widetilde{f}(t_j)]^2$$

(4.93)

where $p_j = 0,1,2, \ldots n$ means a given system of positive numbers. The higher the values of p_j the more effective the smoothing function is and the closer to the given points it lies. We will perform the minimisation $L(f)$ in the domain of cubic splines which, as we know from Ch. 4.3, minimise the first component in formula (4.93). Therefore when inserting $f''(t) = S''(t)$ (4.21) we can represent the functional (4.93) in the form of

$$L(f) = \sum_{j=1}^n \int_{t_{j-1}}^{t_j} \left[m_{j-1} \frac{t_j - t}{h_j} + m_k \frac{t - t_{j-1}}{h_j} \right]^2 dt + \sum_{k=0}^n p_j [y_j - \widetilde{f}(t_j)]^2 .$$

(4.94)

After squaring and calculating the integral, the left-hand side of formula (4.94) is simplified to the form

$$\sum_{j=1}^n \int_{t_{j-1}}^{t_j} \left[m_{j-1} \frac{t_j - t}{h_j} + m_k \frac{t - t_{j-1}}{h_j} \right]^2 dt$$
$$= \sum_{j=1}^n m_j \left[\frac{h_j}{6} m_{j-1} + \frac{h_j + h_{j+1}}{3} m_j + \frac{h_{j+1}}{6} m_{j+1} \right]$$

(4.95)

and as it can be easily proven it can be presented by means of an inner product $(\mathbf{Am}, \mathbf{m})$ in which $\mathbf{Am}$ is given by formula (4.35). Thus we have

$$L(f) = (\mathbf{Am}, \mathbf{m}) + \sum_{k=0}^{n} p_j [y_j - \tilde{y}_j]^2 \tag{4.96}$$

and the minimum of $L(f)$ results from condition

$$\frac{\partial L(f)}{\partial y_j} = 0 . \tag{4.97}$$

Considering (4.30) and performing simple transformations yields

$$\mathbf{H}^T \mathbf{m} + \mathbf{Py} = \mathbf{P}\tilde{\mathbf{y}} \tag{4.98}$$

where $\mathbf{P}$ is the diagonal matrix

$$\mathbf{P} = \begin{bmatrix} p_0 & 0 & 0 & 0 \\ 0 & p_1 & 0 & 0 \\ . & . & . & . \\ 0 & 0 & 0 & p_n \end{bmatrix} . \tag{4.99}$$

Multiplying the left-hand side of (4.98) by $\mathbf{HP}^{-1}$ and considering (4.30) we obtain

$$(\mathbf{A} + \mathbf{HP}^{-1}\mathbf{H}^T)\mathbf{m} = \mathbf{H}\tilde{\mathbf{y}} . \tag{4.100}$$

Solving equation (4.100) with respect to $\mathbf{m}$ gives the vector of the coefficients sought. After inserting it into (4.98) we calculate the values of ordinates

$$\mathbf{y} = \tilde{\mathbf{y}} - \mathbf{P}^{-1}\mathbf{H}^T \mathbf{m} \tag{4.101}$$

which if used in (4.24) permits the approximating smoothing function to be obtained.

4.9. Kalman filter

A popular method of determining a system model with known dynamics given in the state equations domain (2.30), on the base of randomly disturbed response, is the Kalman filter method.

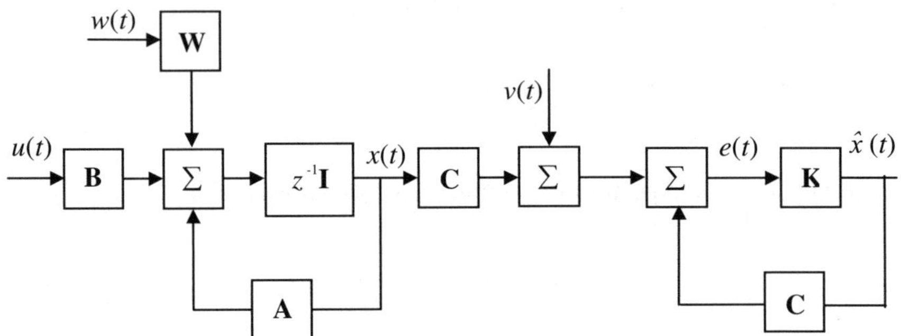

Fig. 4.3. Scheme of the Kalman filter.

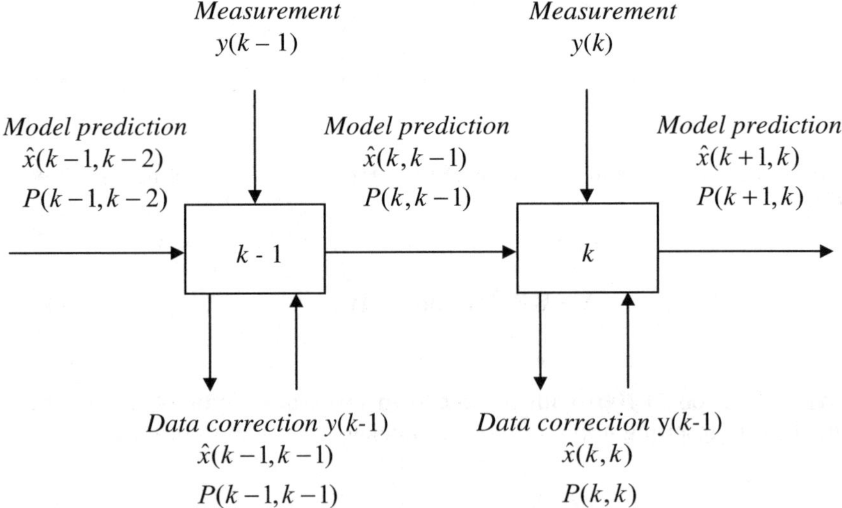

Fig. 4.4. Algorithm of the Kalman filter.

In this method the value of the estimator $\hat{x}(k,k-1)$ is prognosticated in the next step as well as the covariance error $P(k,k-1)$ related to it. The method is based on $(k-1)$ data characterising the output signal, the linear state estimator $\hat{x}(k-1,k-1)$ and the covariances $P(k-1,k-1)$ available at a discrete time moment $(k-1)$. If at the moment (k) the state differs from that expected in the prognosis, then a correction is introduced in the prognosis for the step $(k+1)$ made in step (k). In this way a sequence of estimates $\hat{x}_1, \hat{x}_2, \ldots \hat{x}_n \ldots$ is determined, which should be the best representation of the state $x(k)$ according to the criterion of mean-square error. Fig.4.3. shows a scheme of the Kalman filter and Fig. 4.4 shows the algorithm of its operation [11], [62].

The modified state equation (2.30) considering interferences, is now as follows

$$\begin{aligned} \mathbf{x}(k+1) &= \mathbf{A}(k)\mathbf{x}(k) + \mathbf{B}(k)\mathbf{u}(k) + \mathbf{W}(k)\mathbf{w}(k) \\ \mathbf{y}(k) &= \mathbf{C}(k)\mathbf{x}(k) + \mathbf{v}(k). \end{aligned} \tag{4.102}$$

It is assumed that:

- the deterministic component of the input function $u(t)$ is equal to zero,
- the interferences w and v are mutually uncorrelated,
- v is uncorrelated with x and w,
- the interferences have known matrices of covariances $\mathbf{R}_{ww}$ and $\mathbf{R}_{vv}$.
- the interferences w and v are zero-mean and white

$$\begin{aligned} E[w(k)] &= 0, \\ E[v(k)] &= 0. \end{aligned} \tag{4.103}$$

Estimates of the sought values can be determined based on relations describing the Kalman filter. They are:

State prediction

$$\hat{\mathbf{x}}(k,k-1) = \mathbf{A}(k-1)\hat{\mathbf{x}}(k-1,k-1) + \mathbf{B}(k-1)\mathbf{u}(k-1). \tag{4.104}$$

Covariance prediction

$$\begin{aligned} \mathbf{P}(k,k-1) &= \mathbf{A}(k-1)\mathbf{P}(k-1,k-1)\mathbf{A}^T(k-1) \\ &\quad + \mathbf{W}(k-1)\mathbf{R}_{ww}(k-1)\mathbf{W}^T(k-1). \end{aligned} \tag{4.105}$$

System output prediction

$$\hat{\mathbf{y}}(k,k-1) = \mathbf{C}(k)\hat{\mathbf{x}}(k,k-1) . \tag{4.106}$$

Error of system output prediction

$$\mathbf{e}(k) = \mathbf{y}(k) - \hat{\mathbf{y}}(k,k-1) = \mathbf{y}(k) - \mathbf{C}(k)\hat{\mathbf{x}}(k,k-1) . \tag{4.107}$$

Innovation covariance

$$\mathbf{R}_{ee}(k) = \mathbf{C}(k)\mathbf{P}(k,k-1)\mathbf{C}^{T}(k) + \mathbf{R}_{vv}(k) . \tag{4.108}$$

Matrix of amplification coefficients

$$\mathbf{K}(k) = \mathbf{P}(k,k-1)\mathbf{C}^{T}(k)\mathbf{R}_{ee}^{-1}(k) . \tag{4.109}$$

State correction

$$\hat{\mathbf{x}}(k,k) = \hat{\mathbf{x}}(k,k-1) + \mathbf{K}(k)\mathbf{e}(k) . \tag{4.110}$$

Covariance correction

$$\mathbf{P}(k,k) = [\mathbf{I} - \mathbf{K}(k)\mathbf{C}(k)]\mathbf{P}(k,k-1) . \tag{4.111}$$

Initial conditions: $\mathbf{x}(0)$ has mean $\hat{\mathbf{x}}(0,0)$ and covariance $\mathbf{P}(0,0)$.

The iteration calculations are started by introducing the parameters $\mathbf{A}$, $\mathbf{B}$, $\mathbf{C}$, $\mathbf{P}$, $\mathbf{W}$, the state estimator, and the covariance matrix in the zero step, where most often it is assumed that these are the initial conditions of the state vector $\hat{\mathbf{x}}(0,0)$ and covariance $\mathbf{P}(0,0)$.

4.10. Examples

Example 4.1

Determine the second order Lagrange polynomial that approximates the following measuring data $(y_j, t_j) = (2,1),(4,6),(8,10)$.

Solution

The Lagrange polynomial (4.5) for three approximation points assumes the form

$$
L_N[t,(y_j,t_j)] = y_0 \frac{(t-t_1)(t-t_2)}{(t_0-t_1)(t_0-t_2)}
$$
$$
+ y_1 \frac{(t-t_0)(t-t_2)}{(t_1-t_0)(t_1-t_2)} + y_2 \frac{(t-t_0)(t-t_1)}{(t_2-t_0)(t_2-t_1)} .
$$

$$(4.112)$$

After substituting data and computing we obtain the polynomial sought

$$
L_2(t) = 2 - \frac{1}{15}t + \frac{1}{15}t^2 .
$$

$$(4.113)$$

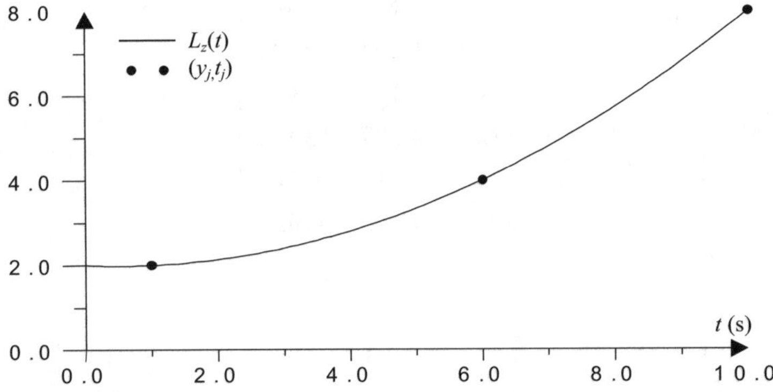

Fig. 4.5. Approximation of measuring data (y_j,t_j) by polynomial $L_2(t)$ (4.113) .

Example 4.2

Applying the least squares method, compute the coefficients a_0, a_1, a_2, a_3, a_4 of the fourth order polynomial $Q(t)$

$$
Q(t) = a_0 + a_1 t + a_2 t^2 + a_3 t^3 + a_4 t^4
$$

$$(4.114)$$

which approximates the set of the following measuring data (y_j, t_j): (0,0);
(0.424, 0.5); (0.663, 1.0); (0.705, 1.5); (0.585, 20); (0.365, 2.5); (0.114, 3.0); (−0.105, 3.5);
(−0.254, 4.0); (−0.314, 4.5).

Solution

The matrices $\mathbf{X}$ and $\mathbf{X}^T$ (4.19) for the given data are as follows

$$\mathbf{X} = \begin{bmatrix} 1 & 0 & 0 & 0 & 0 \\ 1 & 0.5 & 0.25 & 0.12 & 0.06 \\ 1 & 1 & 1 & 1 & 1 \\ 1 & 1.5 & 2.25 & 3.37 & 5.063 \\ 1 & 2 & 4 & 8 & 16 \\ 1 & 2.5 & 6.25 & 15.62 & 39.06 \\ 1 & 3 & 9 & 27 & 81 \\ 1 & 3.5 & 12.25 & 42.87 & 150.06 \\ 1 & 4 & 16 & 64 & 256 \\ 1 & 4.5 & 20.25 & 91.12 & 410.06 \end{bmatrix} \qquad (4.115)$$

$$\mathbf{X}^T = \begin{bmatrix} 1 & 1 & 1 & 1 & 1 & 1 & 1 & 1 & 1 & 1 \\ 0 & 0.5 & 1 & 1.5 & 2 & 2.5 & 3 & 3.5 & 4 & 4.5 \\ 0 & 0.25 & 1 & 2.25 & 4 & 6.25 & 9 & 12.2 & 16 & 20.25 \\ 0 & 0.125 & 1 & 3.37 & 8 & 15.62 & 27 & 42.87 & 64 & 91.12 \\ 0 & 0.063 & 1 & 5.06 & 16 & 39.06 & 81 & 150.06 & 256 & 410.06 \end{bmatrix}. \qquad (4.116)$$

For these matrices the expression $(\mathbf{X}^T\mathbf{X})\mathbf{a} = \mathbf{X}^T\mathbf{y}$ (4.15) has the form

$$\begin{bmatrix} 1 & 22.5 & 71.25 & 253.125 & 958.315 \\ 22.5 & 71.25 & 253.125 & 958.313 & 3.776\cdot10^3 \\ 71.25 & 253.125 & 958.313 & 3.776\cdot10^3 & 1.529\cdot10^4 \\ 253.125 & 958.313 & 3.776\cdot10^3 & 1.529\cdot10^4 & 6.313\cdot10^4 \\ 958.313 & 3.776\cdot10^3 & 1.529\cdot10^4 & 6.313\cdot10^4 & 2.646\cdot10^5 \end{bmatrix} \begin{bmatrix} a_0 \\ a_1 \\ a_2 \\ a_3 \\ a_4 \end{bmatrix}$$

$$= \begin{bmatrix} 1 & 1 & 1 & 1 & 1 & 1 & 1 & 1 & 1 & 1 \\ 0 & 0.5 & 1 & 1.5 & 2 & 2.5 & 3 & 3.5 & 4 & 4.5 \\ 0 & 0.25 & 1 & 2.25 & 4 & 6.25 & 9 & 12.25 & 16 & 20.25 \\ 0 & 0.125 & 1 & 3.375 & 8 & 15.625 & 27 & 42.875 & 64 & 91.125 \\ 0 & 0.063 & 1 & 5.063 & 16 & 39.063 & 81 & 150.063 & 256 & 410.063 \end{bmatrix} \begin{bmatrix} 0 \\ 0.425 \\ 0.663 \\ 0.705 \\ 0.585 \\ 0.365 \\ 0.114 \\ -0.105 \\ -0.254 \\ -0.314 \end{bmatrix} .$$

$$\tag{4.117}$$

In the solution we obtain the sought coefficients, which are: $a_0 = -1.1\cdot10^{-5}$, $a_1 = 1.239$, $a_2 = -0.676$, $a_3 = 0.113$, $a_4 = -6.137\,10^{-3}$. Thus the approximation polynomial $Q(t)$ has the form

$$Q(t) = -1.1\cdot10^{-5} + 1.239\,t - 0.676\,t^2 + 0.113\,t^3 - 6.137\cdot10^{-3}\,t^4 . \tag{4.118}$$

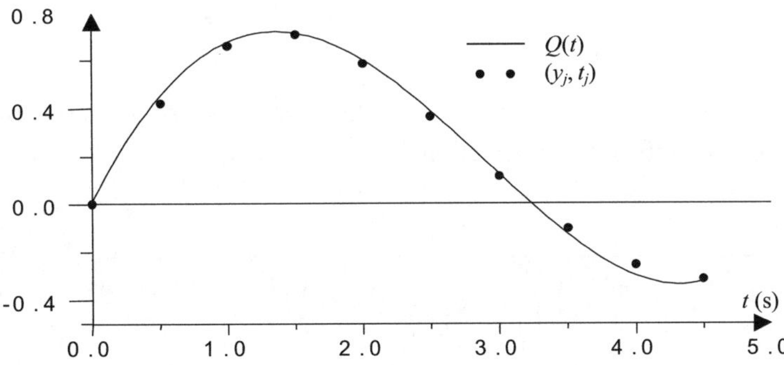

Fig. 4.6. Approximation of measuring data (y_j, t_j) by the polynomial $Q(t)$ (4.118) .

Example 4.3

The square of the magnitude-frequency response has been approximated by the following fraction

$$A^2(\omega) = \frac{1}{\omega^4 - 6\omega + 36}.$$

(4.119)

Determine the differential equation, which corresponds to that response.

Solution

Equating the denominator of the response $A^2(\omega)$ to zero we determine $s_1^2 = 3 + j3\sqrt{3}$, $s_2^2 = 3 - j3\sqrt{3}$ hence the poles are as follows

$$s_{11} = \sqrt{6}\left(\frac{\sqrt{3}}{2} + j\frac{1}{2}\right) \qquad s_{12} = \sqrt{6}\left(-\frac{\sqrt{3}}{2} - j\frac{1}{2}\right)$$

$$s_{21} = \sqrt{6}\left(-\frac{\sqrt{3}}{2} + j\frac{1}{2}\right) \qquad s_{22} = \sqrt{6}\left(\frac{\sqrt{3}}{2} - j\frac{1}{2}\right).$$

(4.120)

In the upper half-plane there are poles s_{11} and s_{21} that, after being substituted into formula (4.40), give

$$K(j\omega) = \frac{1}{\left[\omega - \sqrt{6}\left(\frac{\sqrt{3}}{2} + j\frac{1}{2}\right)\right]\left[\omega - \sqrt{6}\left(-\frac{\sqrt{3}}{2} + j\frac{1}{2}\right)\right]}.$$

(4.121)

Multiplying the denominator of the fraction obtained by j^2 and substituting $s = j\omega$, we obtain the transfer function

$$K(s) = \frac{1}{\left[s - \sqrt{6}\left(j\frac{\sqrt{3}}{2} - \frac{1}{2}\right)\right]\left[s - \sqrt{6}\left(-j\frac{\sqrt{3}}{2} - \frac{1}{2}\right)\right]} = \frac{1}{s^2 + s\sqrt{6} + 6}$$

(4.122)

and corresponding to $K(s)$, the differential equation of the form

$$\frac{d^2 y(t)}{dt^2} + \sqrt{6}\frac{dy(t)}{dt} + 6y(t) = u(t).$$

(4.123)

Example 4.4

Determine the transfer function corresponding to the polynomial $Q(t)$ (4.118), which was determined in example 4.2.

Solution

The polynomial $Q(t)$ has five coefficients c_0 - c_4, which enable the transfer function of the order of n=2 to be determined using the first four coefficients A_0 - A_3 valued:

$$A_0 = c_0 = -1.1 \cdot 10^{-5}, \quad A_1 = c_1 = 1.239, \quad A_2 = 2!c_2 = -1.352, \quad A_3 = 3!c_3 = 0.678.$$

Substituting those coefficients into equation (4.44) we obtain a matrix equation

$$\begin{bmatrix} b_{21} \\ b_{20} \\ a_{21} \\ a_{20} \end{bmatrix} = \begin{bmatrix} 1 & 0 & 0 & 0 \\ 0 & 1 & 1.1 \cdot 10^{-5} & 0 \\ 0 & 0 & -1.239 & 1.1 \cdot 10^{-5} \\ 0 & 0 & 1.325 & -1.239 \end{bmatrix}^{-1} \begin{bmatrix} -1.1 \cdot 10^{-5} \\ 1.239 \\ -1.352 \\ 0.678 \end{bmatrix} \qquad (4.124)$$

whose solution gives: $b_{21} = -1.1 \cdot 10^{-5}$, $b_{20} = 1.239$, $a_{21} = 1.091$, $a_{22} = 0.644$. The sought transfer function has the form

$$K(s) = \frac{-1.1 \cdot 10^{-5} s + 1.239}{s^2 + 1.091s + 0.644}. \qquad (4.125)$$

The impulse response $k(t)$ of transfer function $K(s)$ (4.125) represented in the form of a power series is

$$k(t) = -1.1 \cdot 10^{-5} + 1.238t - 0.6757t^2 + 0.1128t^3 + 5.492 \cdot 10^{-5} t^4 + \dots . \qquad (4.126)$$

It can be easily noted that the first $2n = 4$ coefficients of the series $k(t)$ and $Q(t)$ have the same values and therefore the functions $k(t)$ and $Q(t)$ are identical near the beginning of the time interval. Fig.4.7 shows the results obtained.

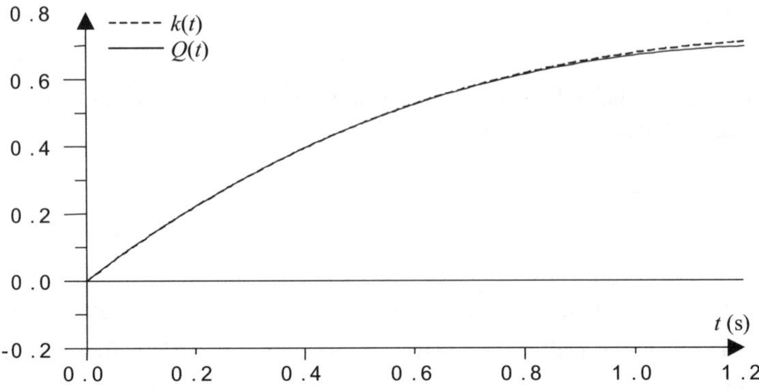

Fig. 4.7. The graphs of impulse response $k(t)$ and the polynomial $Q(t)$ (4.118).

Example 4.5

Using the method of weighted mean and the Nuttall windows, determine the first five coefficients A_k of the expansion in a power series of the following function

$$k(t) = e^{-t} \sin(2t) + e^{-2t} \sin(3t) . \tag{4.127}$$

Solution

Determining the five initial coefficients of the power series requires assuming the fourth power ($p = 4$) of the Nuttall window for which

$$d = \int_{t-\delta}^{t+\delta} \cos^4\left[\frac{\pi}{2\delta}(\tau - t)\right] = \frac{3\delta}{4} \tag{4.128}$$

and substituting the relationships which describe $k(t)$ and d into formulae (4.83) and (4.86) we obtain

$$A_k = \frac{(-1)^k}{k!}\left(\frac{3\delta}{4}\right)^{-1}\int_{t-\delta}^{t+\delta} [e^{-\tau}\sin(2\tau) + e^{-2\tau}\sin(3\tau)]\frac{d^k}{d\tau^k}\cos^4\left(\frac{\pi}{2\delta}(\tau - t)\right) d\tau . \tag{4.129}$$

Computations performed for $t = 0$, $\delta = 0.05$, $k = 0, 1, 2, 3$ and 4 give the following results: $A_0 = -\,0.0016$, $A_1 = 5.007$, $A_2 = -\,7.9928$, $A_3 = 1.1561$, $A_4 = 6.0032$. Comparing the computed values of the coefficients with the first five values of corresponding coefficients of function $k(t)$ (4.127) presented in the form of power series

$$k(t) = 5t - 8t^2 + 1.1667t^3 + 6t^4 - 5.2917t^5 + 1.0889\,t^6 + 0.9379\,t^7 - \ldots \qquad (4.130)$$

it can be easily seen that the results obtained have a very good convergence.

Example 4.6

Using the method of weighted mean and the Nuttall window, determine the first five coefficients A_k of the expansion of function $k(t)$ (4.127) into a power series if it is disturbed by the signal

$$z(t) = 0.1\sin(2000t) + 0.02\sin(3000t). \qquad (4.131)$$

Solution

Making use of the Nuttall window from the previous example and inserting $k_z(t) = k(t) + z(t)$ into formulae (4.88) and (4.92) we obtain

$$A_k = \frac{(-1)^k}{k!}\left(\frac{3\delta}{4}\right)^{-1}\int\limits_{t-\delta}^{t+\delta}\Big[\big[e^{-\tau}\sin(2\tau) + e^{-2\tau}\sin(3\tau) + 0.1\sin(2000\tau) $$

$$ + 0.02\sin(3000\tau)\big]\frac{d^k}{d\tau^k}\cos^4\left(\frac{\pi}{2\delta}(\tau - t)\right)\Big]d\tau. \qquad (4.132)$$

Computations made for $t = 0$, $\delta = 0.05$, $k = 0, 1, 2, 3$ and 4 give the following results $A_0 = -0.0016$, $A_1 = 5.007$, $A_2 = -7.9928$, $A_3 = 1.4535$, $A_4 = 6.0033$ that are identical to those computed in the previous example, with the exception of coefficient A_3, which differs only slightly. This means that in the example considered, the application of the weighted mean caused attenuation of the interferences. The figures included show signal $k(t)$ and interferences $z(t)$ as well as the disturbed signal $k_z(t) = k(t) + z(t)$.

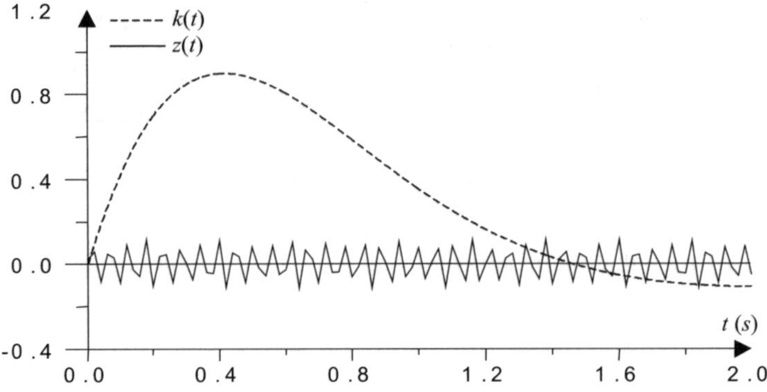

Fig. 4.8. Signal $k(t)$ (4.127) and interference $z(t)$ (4.131) .

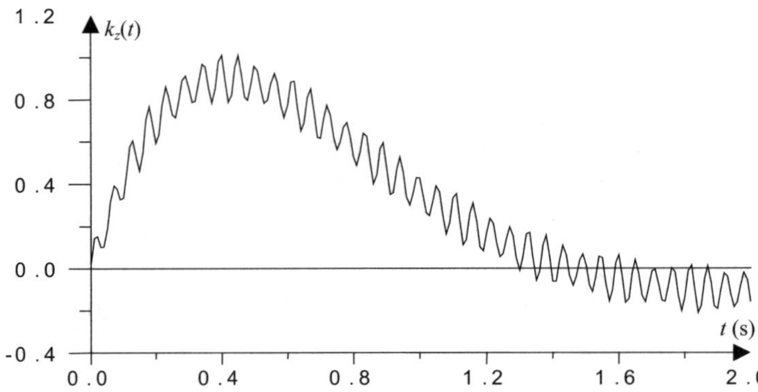

Fig. 4.9. Fig.4.9. Disturbed signal $k_z(t) = k(t) + z(t)$.

Example 4.7

Build four cubic splines corresponding to the measuring data given below

$$\mathbf{t}_j = \begin{bmatrix} t_0 \\ t_1 \\ t_2 \\ t_3 \\ t_4 \end{bmatrix} = \begin{bmatrix} 1 \\ 3 \\ 6 \\ 8 \\ 14 \end{bmatrix} \qquad \mathbf{y}_j = \begin{bmatrix} y_0 \\ y_1 \\ y_2 \\ y_3 \\ y_4 \end{bmatrix} = \begin{bmatrix} 4 \\ 1 \\ 3 \\ 2.5 \\ 4.5 \end{bmatrix}. \tag{4.133}$$

Solution

For the data given in the problem we calculate, using formulae (4.30) - (4.33)

$$\mathbf{h}_j = \begin{bmatrix} h_0 \\ h_1 \\ h_2 \\ h_3 \\ h_4 \end{bmatrix} = \begin{bmatrix} 0 \\ 2 \\ 3 \\ 2 \\ 6 \end{bmatrix} \tag{4.134}$$

$$\mathbf{A} = \begin{bmatrix} \dfrac{h_1 + h_2}{3} & \dfrac{h_2}{6} & 0 \\[2ex] \dfrac{h_2}{6} & \dfrac{h_2 + h_3}{3} & \dfrac{h_3}{6} \\[2ex] 0 & \dfrac{h_3}{6} & \dfrac{h_3 + h_4}{3} \end{bmatrix} = \begin{bmatrix} 1.667 & 0.5 & 0 \\ 0.5 & 1.667 & 0.333 \\ 0 & 0.333 & 2.667 \end{bmatrix} \tag{4.135}$$

$$\mathbf{H} = \begin{bmatrix} \dfrac{1}{h_1} & \left(\dfrac{-1}{h_1} - \dfrac{1}{h_2}\right) & \dfrac{1}{h_2} & 0 & 0 \\[2ex] 0 & \dfrac{1}{h_2} & \left(\dfrac{-1}{h_2} - \dfrac{1}{h_3}\right) & \dfrac{1}{h_3} & 0 \\[2ex] 0 & 0 & \dfrac{1}{h_3} & \left(\dfrac{-1}{h_3} - \dfrac{1}{h_4}\right) & \dfrac{1}{h_4} \end{bmatrix} \tag{4.136}$$

$$= \begin{bmatrix} 0.5 & -0.833 & 0.333 & 0 & 0 \\ 0 & 0.333 & -0.833 & 0.5 & 0 \\ 0 & 0 & 0.5 & -0.667 & 0.167 \end{bmatrix}$$

$$\mathbf{m} = \mathbf{A}^{-1}\mathbf{Hy} = \begin{bmatrix} m_0 \\ m_1 \\ m_2 \\ m_3 \\ m_4 \end{bmatrix} = \begin{bmatrix} 0 \\ 1.633 \\ -1.112 \\ 0.358 \\ 0 \end{bmatrix} \tag{4.137}$$

which, after inserting into the formula for $S(t)$ (4.22), and carrying out calculations for $j = 1, 2, 3$ and 4 gives:

$$\begin{aligned}
&for \quad j = 1 \quad t \in [1,3] \\
&S_1(t) = 0.1363\,t^3 - 0.4091 t^2 - 1.6350\,t + 5.9078 \\
&for \quad j = 2 \quad t \in [3,6] \\
&S_2(t) = -0.1525\,t^3 + 2.1889\,t^2 - 9.4272\,t + 13.699 \\
&for \quad j = 3 \quad t \in [6,8] \\
&S_3(t) = 0.1226\,t^3 - 2.7622\,t^2 + 20.281\,t - 45.723 \\
&for \quad j = 4 \quad t \in [8,14] \\
&S_4(t) = -0.0099\,t^3 + 0.4173\,t^2 - 5.1516\,t + 22.091.
\end{aligned} \tag{4.138}$$

The diagram in Fig.4.10 shows splines $S_1(t)$ - $S_4(t)$.

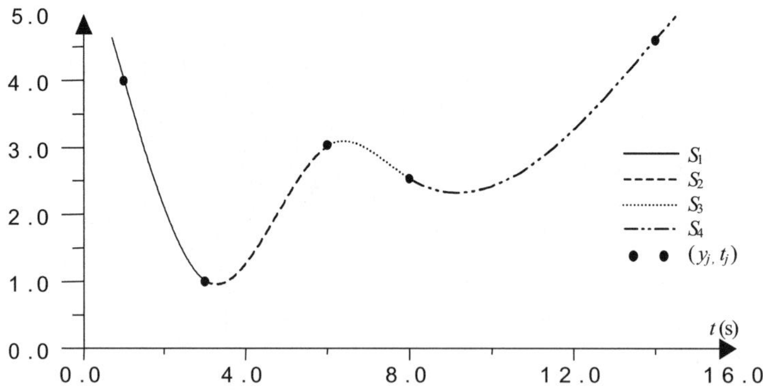

Fig.4.10. Approximation of measuring data (y_j, t_j) (4.133) by the cubic splines $S_1(t)$ - $S_4(t)$ (4.138).

Example 4.8

Build a smoothing function, assuming the data from the previous problem was disturbed, and is now

$$\mathbf{t}_j = \begin{bmatrix} t_0 \\ t_1 \\ t_2 \\ t_3 \\ t_4 \end{bmatrix} = \begin{bmatrix} 1 \\ 3 \\ 6 \\ 8 \\ 14 \end{bmatrix} \qquad \tilde{\mathbf{y}}_j = \begin{bmatrix} \tilde{y}_0 \\ \tilde{y}_1 \\ \tilde{y}_2 \\ \tilde{y}_3 \\ \tilde{y}_4 \end{bmatrix} = \begin{bmatrix} 3.9 \\ 1.1 \\ 2.95 \\ 2.6 \\ 4.45 \end{bmatrix}. \tag{4.139}$$

Solution

Examining the data in this problem reveals that matrices $\mathbf{A}$, $\mathbf{H}$ and vector $\mathbf{h}$ are identical to those in example 4.7. In order to calculate the coefficients $\mathbf{m}$ of the smoothing function let us assume as an example, that the diagonal matrix $\mathbf{P}$ has the form

$$\mathbf{P} = \begin{bmatrix} 20 & 0 & 0 & 0 & 0 \\ 0 & 20 & 0 & 0 & 0 \\ 0 & 0 & 20 & 0 & 0 \\ 0 & 0 & 0 & 20 & 0 \\ 0 & 0 & 0 & 0 & 20 \end{bmatrix} \tag{4.140}$$

and let us mark the left-hand side of equation (4.100) as $\mathbf{C}$

$$\mathbf{C} = (\mathbf{A} + \mathbf{H}\mathbf{P}^{-1}\mathbf{H}^T) = \begin{bmatrix} 1.719 & 0.472 & 8.33\cdot10^{-3} \\ 0.472 & 1.719 & 0.296 \\ 8.33\cdot10^{-3} & 0.296 & 2.703 \end{bmatrix}. \tag{4.141}$$

We now calculate the coefficients $\mathbf{m}$ of the smoothing function from formula (4.100)

$$\mathbf{m} = \mathbf{C}^{-1}\mathbf{H}\tilde{\mathbf{y}} = \begin{bmatrix} 1.418 \\ -0.897 \\ 0.273 \end{bmatrix}. \tag{4.142}$$

Inserting **m** (4.142) into (4.101) we calculate the values of the ordinates of this function

$$\mathbf{y}_j = \tilde{\mathbf{y}} - \mathbf{P}^{-1}\mathbf{H}^T\mathbf{m} = \begin{bmatrix} 3.865 \\ 1.174 \\ 2.906 \\ 2.632 \\ 4.448 \end{bmatrix}. \tag{4.143}$$

On the base of coefficients

$$\mathbf{m}_j = \begin{bmatrix} m_0 \\ m_1 \\ m_2 \\ m_3 \\ m_4 \end{bmatrix} = \begin{bmatrix} 0 \\ 1.418 \\ -0.897 \\ 0.273 \\ 0 \end{bmatrix} \tag{4.144}$$

and of the coordinates $\mathbf{y}_j$ (4.143) we determine function $S(t)$ in the same way as in problem 4.7 by making use of formula (4.22). Having made calculations for $j = 1, 2, 3$ and 4 we obtain the following functions

$$\begin{aligned}
&\textit{for} \quad j = 1 \quad t \in [1,3] \\
&S_1(t) = 0.1182\,t^3 - 0.3547\,t^2 - 1.4631\,t + 5.5641 \\
&\textit{for} \quad j = 2 \quad t \in [3,6] \\
&S_2(t) = -0.1286\,t^3 + 1.8666\,t^2 - 8.1268\,t + 12.228 \\
&\textit{for} \quad j = 3 \quad t \in [6,8] \\
&S_3(t) = 0.0975\,t^3 - 2.2028\,t^2 + 16.2901\,t - 36.606 \\
&\textit{for} \quad j = 4 \quad t \in [8,14] \\
&S_4(t) = -0.0076\,t^3 + 0.3179\,t^2 - 3.8762\,t + 17.168.
\end{aligned} \tag{4.145}$$

The diagram in Fig. 4.11 shows the smoothing function being sought.

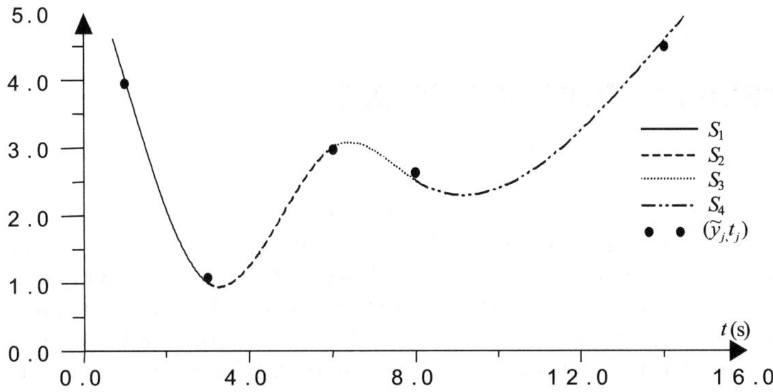

Fig. 4.11. Approximation of measuring data $(\widetilde{y}_j, t_j)$ (4.139) by smoothing functions $S_1(t)$-$S_4(t)$ (4.145) .

5. SIMPLIFICATION OF MODELS

The principal purpose of simplifying a model is to obtain it in a less complicated form, which makes the analysis of its properties easier, faster and cheaper. The simplification procedure leads either to lowering the model order, usually connected with changes in the model parameters, or to adopting a different form of the model with its number of parameters reduced. It is assumed that the simplified models have to fulfil a chosen objective function, which determines the application of the corresponding simplification method. The most popular objective functions deal with

- the difference between model responses with the integral-square-error criterion,

- the consistency of derivatives of model responses at the beginning of the time, or

- the consistency of model responses at a determined interval of time.

In cases when the postulate of minimum differences of the responses is not posed, the models are being simplified based on analyses of the position of the eigenvalues and mutual relations occurring between them. The eigenvalues are reduced depending on the value of their negative real part. Eigenvalues with small negative real parts are left behind in the model if the mapping error has to be small for longer time periods. Eigenvalues with high negative real parts are left in the model, reducing the remaining ones, if the model has to map the primary one with a small error for short time periods. However it is worth emphasising that the choice of the simplification method is not always free, as often it can be imposed by the form of the primary model. For example if the primary model is in the form of an algebraic polynomial and we want to reduce its order, then the choice of criterion for searching the simplified model is for obvious reasons restricted. The choice is then limited to minimising the distance between the polynomials or for instance to adopting the criterion on consistency of the polynomial derivatives at zero. If the primary model is given in the form of a state equation or transfer function then the question of criterion choice, according to which it will be simplified, is as a rule, arbitrary. This chapter is devoted to the discussion of selected methods of model simplification and presentation of the necessary computation procedures, which lead to obtaining such models.

5.1. The least-squares approximation

5.1.1. Orthogonal polynomials

In the previous chapter the problem of making the least-square approximation fit the collection of data, was considered. The other approximation problem considered below will address the approximation of functions. Suppose then, that the mathematical model of a system $f(t) \in [a,b]$ is given, and that it is represented by a high order polynomial. Our task is to determine a polynomial of a lower order which will approximate $f(t)$ with a minimum integral-square-error. Suppose additionally that it will be an orthogonal polynomial. In this way our task is reduced to determining the real coefficients $a_0, a_1, \dots a_n$ that will minimise the functional

$$I_2(a_0, a_1, \dots a_n) = \min_{a_k} \int_a^b w(t) \left(f(t) - \sum_{k=0}^n a_k \Phi_k(t) \right)^2 dt \tag{5.1}$$

where $w(t)$ is weight function which orthogonalises the polynomial $\Phi_k(t)$ over the interval $[a, b]$. Minimising $I_2(a_0, a_1, \dots a_n)$ (5.1) requires that the derivatives for every $j = 0, 1, \dots n$ become equal to zero

$$\frac{\partial I_2}{\partial a_j} = 2 \int_a^b w(t) \left[f(t) - \sum_{k=0}^n a_k \Phi_k(t) \right] \Phi_j(t) \, dt = 0 \qquad j = 0, 1, \dots n. \tag{5.2}$$

Let us write the equation system (5.2) in the normal form

$$\int_a^b w(t) f(t) \Phi_j(t) \, dt = \sum_{k=0}^n a_k \int_a^b w(t) \Phi_k(t) \Phi_j(t) \, dt. \tag{5.3}$$

From the condition of orthogonality, it results that the right hand side of equation (5.3) takes the values

$$\int_a^b w(t) \Phi_k(t) \Phi_j(t) \, dt = \begin{cases} 0 & for \quad j \neq k \\ \alpha_k > 0 & for \quad j = k. \end{cases} \tag{5.4}$$

If in (5.4) $\alpha_k = 1$ for each $k = 0, 1, \ldots n$ then $\Phi_k(t)$ are called orthonormal polynomials. They can be obtained by dividing $\Phi_k(t)$ by $\|\Phi_k(t)\|$ where

$$\|\Phi_k(t)\| = \sqrt{\int_a^b \Phi_k(t)\Phi_k(t)\,dt} \ . \tag{5.5}$$

Taking into account relationship (5.4) in formula (5.3), it is easy to determine the coefficients a_k being sought

$$a_k = \frac{\displaystyle\int_a^b w(t)f(t)\Phi_k(t)dt}{\displaystyle\int_a^b w(t)\Phi_k^2(t)dt} = \frac{1}{\alpha_k}\int_a^b w(t)f(t)\Phi_k(t)dt \quad k = 1, 2, \ldots n \tag{5.6}$$

where in the case of orthogonal $\Phi_k(t)$ the denominator $\alpha_k = 1$. Hence, if $\{\Phi_0, \Phi_0 \ldots \Phi_n\}$ is a set of orthogonal polynomials over the interval $[a, b]$ with the weight function $w(t)$ then the polynomial

$$\sum_{k=0}^n a_k \Phi_k(t) \tag{5.7}$$

approximates function $f(t)$ over the interval $[a,b]$ giving a minimum of the integral-square-error. The set of polynomials $\{\Phi_0, \Phi_1 \ldots \Phi_n\}$ is always orthogonal over $[a,b]$ with the weight function $w(t)$, if it is defined in the following way [10]:

$$\Phi_0(t) = 1 , \tag{5.8}$$

$$\Phi_1(t) = t - B_1$$

$$B_1 = \frac{\displaystyle\int_a^b t\,w(t)[\Phi_0(t)]^2\,dt}{\displaystyle\int_a^b w(t)[\Phi_0]^2(t)dt} \tag{5.9}$$

and for $k \geq 2$

$$\Phi_k(t) = (t - B_k)\Phi_{k-1}(t) - C_k)\Phi_{k-2}(t)$$

$$B_k = \frac{\displaystyle\int_a^b t\, w(t)\,[\Phi_{k-1}(t)]^2\, dt}{\displaystyle\int_a^b w(t)\,[\Phi_{k-1}]^2(t)\, dt}$$

$$C_k = \frac{\displaystyle\int_a^b t\, w(t)\,\Phi_{k-1}(t)\Phi_{k-2}(t)\, dt}{\displaystyle\int_a^b w(t)\,[\Phi_{k-2}]^2(t)\, dt}\,.$$

(5.10)

5.1.2. Chebyshev polynomials

One of the most common sets of orthogonal polynomials is the set of Chebyshev polynomials with a weight function in the form

$$w(z) = \frac{1}{\sqrt{1-z^2}}\,.$$

(5.11)

Let us consider the Chebyshev polynomials

$$T_k(z) = \cos(k\, arc\cos z) \qquad k \geq 0 \qquad -1 \leq z \leq 1.$$

(5.12)

Substituting

$$\theta = arc\cos z$$

(5.13)

changes equation (5.12) to

$$T_k(\theta) = \cos(k\theta) \qquad \theta \in [0,\pi]\,.$$

(5.14)

Thus we have

$$T_{k+1}(\theta) = \cos[(k+1)\theta] = \cos(k\theta)\cos\theta - \sin(k\theta)\sin\theta$$

(5.15)

and

$$T_{k-1}(\theta) = \cos[(k-1)\theta] = \cos(k\theta)\cos\theta + \sin(k\theta)\sin\theta \qquad (5.16)$$

whence results the simple relationship

$$T_{k+1}(\theta) = 2\cos(k\theta)\cos\theta - T_{k-1}(\theta). \qquad (5.17)$$

Returning to the variable z we obtain

$$T_{k+1}(z) = 2zT_k(z) - T_{k-1}(z) \qquad k \geq 1 \qquad (5.18)$$

where

$$\begin{aligned} T_0(z) &= \cos(0 \cdot arc\cos z) = 1 \\ T_1(z) &= \cos(1 \cdot arc\cos z) = z \end{aligned} \qquad (5.19)$$

while the consecutive polynomials that result from (5.18) are

$$\begin{aligned} T_2(z) &= 2zT_1(z) - T_0(z) = 2z^2 - 1 \\ T_3(z) &= 2zT_2(z) - T_1(z) = 4z^3 - 3z \\ T_4(z) &= 2zT_3(z) - T_2(z) = 8z^4 - 8z^2 + 1 \end{aligned} \qquad (5.20)$$

....

etc.

It can be easily shown that those polynomials are orthogonal over the interval $[-1,1]$ with a weight function (5.11). For we have [10]:

$$\int_{-1}^{1} \frac{T_k(z)T_j(z)}{\sqrt{1-z^2}}\, dz = \int_{-1}^{1} \frac{\cos(k\, arc\cos z)\cos(j\, arc\cos z)}{\sqrt{1-z^2}}\, dz. \qquad (5.21)$$

Reintroducing substitution (5.13) we have

$$\int_{-1}^{1}\frac{T_k(z)T_j(z)}{\sqrt{1-z^2}}\,dz = \int_{\pi}^{0}\frac{\cos(k\theta)\cos(j\theta)}{\sin\theta}(-\sin\theta)d\theta$$

$$= \int_{\pi}^{0}\cos(k\theta)\cos(j\theta)\,d\theta = 0 \qquad for \qquad j\neq k. \tag{5.22}$$

In a similar way it can be also shown that

$$\int_{-1}^{1}\frac{T_k(z)T_j(z)}{\sqrt{1-z^2}}\,dz = \int_{-1}^{1}\frac{[T_k(z)]^2}{\sqrt{1-z^2}}\,dz = \frac{\pi}{2} \qquad for \qquad j=k. \tag{5.23}$$

By replacing the variable z from the interval $[-1,1]$ with the variable of time t from the interval $[0, t_f]$ according to the relationship

$$z = 1 - 2\frac{t}{t_f} \qquad 0 \le t \le t_f \tag{5.24}$$

the following first Chebyshev polynomials $T_k(t)$ over $[0, t_f]$ are obtained

$$T_0(t) = 1$$

$$T_1(t) = 1 - 2\left(\frac{t}{t_f}\right)$$

$$T_2(t) = 1 - 8\left(\frac{t}{t_f}\right) + 8\left(\frac{t}{t_f}\right)^2$$

$$T_3(t) = 1 - 18\left(\frac{t}{t_f}\right) + 48\left(\frac{t}{t_f}\right)^2 - 32\left(\frac{t}{t_f}\right)^3 \tag{5.25}$$

$$T_4(t) = 1 - 32\left(\frac{t}{t_f}\right) + 160\left(\frac{t}{t_f}\right)^2 - 256\left(\frac{t}{t_f}\right)^3 + 128\left(\frac{t}{t_f}\right)^4$$

$$\cdots\cdots\cdots$$

$$T_{k+1}(t) = \left(2 - \frac{4t}{t_f}\right)T_k(t) - T_{k-1}(t).$$

It can be easily proven that the orthogonality of the $T_k(t)$ polynomials (5.25) is obtained for the weight function $w(t)$ in the form

$$w(t) = \frac{1}{\sqrt{t_f\, t - t^2}}$$

(5.26)

for which

$$\int_0^{t_f} w(t)T_k(t)T_j(t)dt = \begin{cases} 0 & j \neq k \\ \dfrac{\pi}{2} & j = k \neq 0 \\ \pi & j = k = 0. \end{cases}$$

(5.27)

The coefficients of the Chebyshev polynomial a_k that approximate function $f(t)$ over the interval $[0,\, t_f]$ with the minimum value of the integral-square-error, are determined based on (5.6) with (5.26) and (5.27) taken into consideration. As a result we obtain

$$a_0 = \frac{1}{\pi}\int_0^{t_f} w(t)f(t)T_0(t)dt$$

$$a_k = \frac{2}{\pi}\int_0^{t_f} w(t)f(t)T_k(t)dt \qquad k = 0,1,\ \dots\ n.$$

(5.28)

The approximation polynomial of Chebyshev has thus the form

$$T_n(t) = \sum_{k=0}^{n} a_k T_k(t)$$

(5.29)

where T_k and a_k are determined by relationships (5.25) and (5.28) respectively.

5.1.3. Legendre polynomials

Another set of orthogonal polynomials is the set of Legendre polynomials $\{\Phi_0, \Phi_1,\ \dots\ \Phi_n\}$ over the interval $[-1,1]$ that minimise functional (5.1) and satisfy the following conditions:

$$w(t) \equiv 1$$
$$\Phi_0(t) \equiv 1. \tag{5.30}$$

Using formulae (5.8) - (5.10) we now obtain

$$\Phi_1(t) = t - B_1$$

$$B_1 = \frac{\int\limits_{-1}^{1} t\, dt}{\int\limits_{-1}^{1} dt} = 0. \tag{5.31}$$

$$\Phi_2(t) = (t - B_2)\Phi_1(t) - C_2\Phi_0(t) = t^2 - \frac{1}{3}$$

$$B_2 = \frac{\int\limits_{-1}^{1} t^3\, dt}{\int\limits_{-1}^{1} t^2\, dt} = 0$$

$$C_2 = \frac{\int\limits_{-1}^{1} t^2\, dt}{\int\limits_{-1}^{1} dt} = \frac{1}{3}. \tag{5.32}$$

Computing in the same way, we obtain

$$\Phi_3(t) = t^3 - \frac{3}{5}t$$

$$\Phi_4(t) = t^4 - \frac{6}{7}t^2 + \frac{3}{35}$$

$$\Phi_5(t) = t^5 - \frac{10}{9}t^3 + \frac{5}{21}t \tag{5.33}$$

$$\cdots\cdots$$

etc.

The polynomials (5.31) - (5.33) are called Legendre polynomials. The a_k coefficients of those polynomials, approximating function $f(t)$ over the interval $[-1,1]$ with the minimum value of the integral-square-error, can be determined in identically the same way as the Chebyshev polynomials, using formula (5.6).

5.2. The Rao-Lamba method

The simplification of models in the frequency domain is reduced to the parametric optimisation of the transfer function coefficients, in such a way so that the value of the chosen error criterion is minimised. The equivalent criterion of error resulting from the Parseval theorem and given by

$$I_2(\omega) = \frac{1}{2\pi j} \int_{-j\infty}^{+j\infty} E(s)E(-s)\,ds \tag{5.34}$$

where $E(s)$ is the error Laplace transform, corresponds in the frequency domain to the integral-square-error criterion in the time domain. Inserting $s = j\omega$ into (5.34) and adopting the lower integration limit equal to zero at double the value of the integral, we obtain

$$I_2(\omega) = \frac{1}{\pi} \int_0^\infty E(j\omega)E(-j\omega)\,d\omega = \frac{1}{\pi} \int_0^\infty |E(j\omega)|^2 d\omega. \tag{5.35}$$

On the basis of formula (5.35) and presenting $E(\omega)$ as a function of the parameters of the simplified model, the optimisation is performed in such a way so that a minimum of $I_2(\omega)$ is obtained. In practice, optimisation is carried out over finite integration limits between $\omega_{\min}$ and $\omega_{\max}$ which reduces our problem to minimisation

$$I_2(\omega) = \min \int_{\omega_{\min}}^{\omega_{\max}} |E(j\omega)|^2 \, d\omega. \tag{5.36}$$

Let us represent the n-order primary model in the form of spectral transmittance

$$\frac{Y_n(j\omega)}{U_n(j\omega)} = \frac{b_m(j\omega)^m + b_{m-1}(j\omega)^{m-1} + ... + b_1(j\omega) + b_0}{(j\omega)^n + a_{n-1}(j\omega)^{n-1} + ...a_1(j\omega) + a_0} = \frac{B(j\omega)}{A(j\omega)} \qquad m < n \tag{5.37}$$

and also the simplified model in a similar form, assuming that it is of the order of q, where $q < n$

$$\frac{Y_q(j\omega)}{U_q(j\omega)} = \frac{d_p(j\omega)^p + d_{p-1}(j\omega)^{p-1} + ... + d_1(j\omega) + d_0}{(j\omega)^q + c_{q-1}(j\omega)^{q-1} + ...c_1(j\omega) + c_0} = \frac{D(j\omega)}{C(j\omega)} \quad p < q. \quad (5.38)$$

For these models, assuming a common input function, the minimisation of the functional (5.36) is reduced to minimisation [78], [86]:

$$I_2(\omega) = \min \int_{\omega_{min}}^{\omega_{max}} |B(j\omega)C(j\omega) - A(j\omega)D(j\omega)|^2 \, d\omega. \quad (5.39)$$

Minimum $I_2(\omega)$ (5.39) is achieved from the conditions

$$\frac{\partial I_2(\omega)}{\partial d_i} = 0 \qquad i = 0,1, ... \, p$$

$$\frac{\partial I_2(\omega)}{\partial c_i} = 0 \qquad i = 0,1, ... \, q-1$$

$$(5.40)$$

which leads to a linear equation system, that after being solved, gives the values of the coefficients being sought for the simplified model.

5.3. Criterion of consistency of model response derivatives at the origin

The accuracy of mapping models at the beginning of the time interval is of particular importance in systems whose rated operation regime is a dynamic state. In principle, the responses of such systems in the time domain do not reach a steady state. Measuring systems that co-operate with piezoelectric transducers, capacitive transducers and similar ones applied in dynamic metrology domain are good examples of this. The criterion of agreement of response derivatives at the origin requires the coefficients

$$c_{n,k} = c_{q,k} \qquad k = 0,1,2, ... \quad (5.41)$$

of the n-order primary model $k_n(t)$ impulse response expanded in a series

$$k_n(t) = \sum_{k=0}^{\infty} \frac{1}{k!} A_{n,k}\, t^k = \sum_{k=0}^{\infty} c_{n,k} t^k \tag{5.42}$$

to be equal to the coefficients of the simplified q-order model $k_q(t)$

$$k_q(t) = \sum_{k=0}^{\infty} \frac{1}{k!} A_{q,k} t^k = \sum_{k=0}^{\infty} c_{q,k} t^k \qquad q < n \,. \tag{5.43}$$

The postulate (5.41) means that the impulse responses of models $k_n(t)$ and $k_q(t)$ together with their k derivatives have identical initial conditions at zero, that is that the simplified model will represent the primary model with a negligible error over the initial time interval. In order to determine a model that satisfies condition (5.41) up to $2q$ terms, let us note that an explicit relationship exists (4.42) between the parameters $a_{n,0}, a_{n,1}, \ldots a_{n,n-1}$ and $b_{n,0}, b_{n,1}, \ldots b_{n,m}$, of the n-order model (2.24) and the coefficients of its impulse response in the form of the series (5.42). It permits the model to be directly determined if only the responses $k_n(t)$ and their derivatives at zero are known. Using equations (4.42) and (4.50) we can easily determine the simplified model that satisfies condition (5.41), in the three following steps:

- for the n-order of the primary model we compute the coefficients $A_{n,0}$, $A_{n,1}$, ... $A_{n,n-1}$ of its impulse response, represented in the form of a power series, according to formula (4.50),
- we assume the order q of the simplified model $q < n$,
- we compute the coefficients of the numerator and denominator of the simplified model according to formula (4.42), taking into account only the $2q$ first coefficients $A_{n,0}$, $A_{n,1}$, ... $A_{n,n-1}$ calculated in the first step.

The simplified model of the order of q determined in such a way, has the property that the first $2q$ coefficients of its impulse response, represented in the form of a power series, are identical with the first $2q$ coefficients of the primary model. This means a very accurate mapping of models over the initial time interval.

5.4. Reduction of state matrix order with selected eigenvalues retained

The method allows the order of state matrix $\mathbf{A}_{(nxn)}$ in equation

$$\dot{\mathbf{x}}(t) = \mathbf{A}\,\mathbf{x}(t) + \mathbf{B}\,u(t) \tag{5.44}$$

to be lowered in such a way so that in the reduced matrix $\mathbf{A}_{r(l \times l)}$

$$\dot{\mathbf{x}}(t) = \mathbf{A}_r \mathbf{x}(t) + \mathbf{B}_r u(t) \qquad \text{where} \qquad l < n \tag{5.45}$$

some selected eigenvalues of matrix $\mathbf{A}$ would be retained [15]. Retaining high negative eigenvalues in matrix $\mathbf{A}_r$ and the elimination of small negative eigenvalues renders an effect which is similar to that obtained by the method previously presented, in that there is a negligible mapping error for short time intervals. However, retaining low negative eigenvalues in matrix $\mathbf{A}_r$ with the remaining ones being omitted, gives a low mapping error for longer time intervals and a higher error in dynamic states near the origin. This is due to the fact that the neglected eigenvalues make a very insignificant contribution to the total response except at the beginning of the time interval. In order to determine the reduced matrices $\mathbf{A}_r$ let us consider a simple example in which matrices $\mathbf{A}$ and $\mathbf{B}$ of state equation (5.44) have the form

$$\mathbf{A} = \begin{bmatrix} a_{1,1} & a_{1,2} & a_{1,3} & a_{1,4} \\ a_{2,1} & a_{2,2} & a_{2,3} & a_{2,4} \\ a_{3,1} & a_{3,2} & a_{3,3} & a_{3,4} \\ a_{4,1} & a_{4,2} & a_{4,3} & a_{4,4} \end{bmatrix} \qquad \mathbf{B} = \begin{bmatrix} b_1 \\ b_2 \\ b_3 \\ b_4 \end{bmatrix} \quad \text{and} \quad u(t) = \mathbf{1}(t) \tag{5.46}$$

and the modal matrix $\mathbf{S}$ of matrix $\mathbf{A}$ and its inverse $\mathbf{S}^{-1}$ are

$$\mathbf{S} = \begin{bmatrix} x_{1,1} & x_{1,2} & x_{1,3} & x_{1,4} \\ x_{2,1} & x_{2,2} & x_{2,3} & x_{2,4} \\ x_{3,1} & x_{3,2} & x_{3,3} & x_{3,4} \\ x_{4,1} & x_{4,2} & x_{4,3} & x_{4,4} \end{bmatrix} \qquad \mathbf{S}^{-1} = \begin{bmatrix} \varphi_{1,1} & \varphi_{1,2} & \varphi_{1,3} & \varphi_{1,4} \\ \varphi_{2,1} & \varphi_{2,2} & \varphi_{2,3} & \varphi_{2,4} \\ \varphi_{3,1} & \varphi_{3,2} & \varphi_{3,3} & \varphi_{3,4} \\ \varphi_{4,1} & \varphi_{4,2} & \varphi_{4,3} & \varphi_{4,4} \end{bmatrix} \tag{5.47}$$

where the eigenvectors of $\mathbf{S}$ are normalised

$$\sum_{i=1}^{4} \left(x_{i,j} \right)^2 = 1 \qquad j = 1, 2, 3, 4 \, . \tag{5.48}$$

Let us write the solution $\mathbf{x}(t)$ of the state equation (5.44) in the following way

$$\begin{bmatrix} x_1 \\ x_2 \\ x_3 \\ x_4 \end{bmatrix} = \int\limits_0^t \begin{bmatrix} x_{1,1} & x_{1,2} & x_{1,3} & x_{1,4} \\ x_{2,1} & x_{2,2} & x_{2,3} & x_{2,4} \\ x_{3,1} & x_{3,2} & x_{3,3} & x_{3,4} \\ x_{4,1} & x_{4,2} & x_{4,3} & x_{4,4} \end{bmatrix} \begin{bmatrix} e^{\lambda_1(t-\tau)} & 0 & 0 & 0 \\ 0 & e^{\lambda_2(t-\tau)} & 0 & 0 \\ 0 & 0 & e^{\lambda_3(t-\tau)} & 0 \\ 0 & 0 & 0 & e^{\lambda_4(t-\tau)} \end{bmatrix}$$

$$\cdot \begin{bmatrix} \varphi_{1,1} & \varphi_{1,2} & \varphi_{1,3} & \varphi_{1,4} \\ \varphi_{2,1} & \varphi_{2,2} & \varphi_{2,3} & \varphi_{2,4} \\ \varphi_{3,1} & \varphi_{3,2} & \varphi_{3,3} & \varphi_{3,4} \\ \varphi_{4,1} & \varphi_{4,2} & \varphi_{4,3} & \varphi_{4,4} \end{bmatrix} \begin{bmatrix} b_1 \\ b_2 \\ b_3 \\ b_4 \end{bmatrix} d\tau. \tag{5.49}$$

After simple transformations, equation (5.49) can be represented in the form

$$\begin{bmatrix} x_1 \\ x_2 \\ x_3 \\ x_4 \end{bmatrix} = \sum_{i=1}^{4} \frac{-1+e^{\lambda_i t}}{\lambda_i} \begin{bmatrix} x_{1,i} \\ x_{2,i} \\ x_{3,i} \\ x_{4,i} \end{bmatrix} (\varphi_{i,1}b_1 + \varphi_{i,2}b_2 + \varphi_{i,3}b_3 + \varphi_{i,4}b_4). \tag{5.50}$$

As an example, suppose we want to reduce equation (5.50) in such a way so that matrix $\mathbf{A}_r$ would retain the two first eigenvalues λ_1, λ_2 while the eigenvalues λ_3 and λ_4 would be eliminated. For such an assumption, equation (5.50) is reduced to the form

$$\begin{bmatrix} x_1 \\ x_2 \\ x_3 \\ x_4 \end{bmatrix} = \sum_{i=1}^{2} \frac{-1+e^{\lambda_i t}}{\lambda_i} \begin{bmatrix} x_{1,i} \\ x_{2,i} \\ x_{3,i} \\ x_{4,i} \end{bmatrix} (\varphi_{i,1}b_1 + \varphi_{i,2}b_2 + \varphi_{i,3}b_3 + \varphi_{i,4}b_4). \tag{5.51}$$

If we denote

$$\xi_i = \frac{-1+e^{\lambda_i t}}{\lambda_i} (\varphi_{i,1}b_1 + \varphi_{i,2}b_2 + \varphi_{i,3}b_3 + \varphi_{i,4}b_4) \tag{5.52}$$

then for the two first variables x_1 and x_2 equation (5.51) assumes the form

$$\begin{bmatrix} x_1 \\ x_2 \end{bmatrix} = \begin{bmatrix} x_{1,1} & x_{1,2} \\ x_{2,1} & x_{2,2} \end{bmatrix} \begin{bmatrix} \xi_1 \\ \xi_2 \end{bmatrix} \tag{5.53}$$

and a similar one for the two remaining variables x_3 and x_4

$$\begin{bmatrix} x_3 \\ x_4 \end{bmatrix} = \begin{bmatrix} x_{3,1} & x_{3,2} \\ x_{4,1} & x_{4,2} \end{bmatrix} \begin{bmatrix} \xi_1 \\ \xi_2 \end{bmatrix}. \tag{5.54}$$

Substituting the solution

$$\begin{bmatrix} \xi_1 \\ \xi_2 \end{bmatrix} = \begin{bmatrix} x_{1,1} & x_{1,2} \\ x_{2,1} & x_{2,2} \end{bmatrix}^{-1} \begin{bmatrix} x_1 \\ x_2 \end{bmatrix} \tag{5.55}$$

in equation (5.54) we obtain

$$\begin{bmatrix} x_3 \\ x_4 \end{bmatrix} = \begin{bmatrix} x_{3,1} & x_{3,2} \\ x_{4,1} & x_{4,2} \end{bmatrix} \begin{bmatrix} x_{1,1} & x_{1,2} \\ x_{2,1} & x_{2,2} \end{bmatrix}^{-1} \begin{bmatrix} x_1 \\ x_2 \end{bmatrix}. \tag{5.56}$$

Let us now write the two first derivatives of expression

$$\begin{bmatrix} \dot{x}_1 \\ \dot{x}_2 \\ \dot{x}_3 \\ \dot{x}_4 \end{bmatrix} = \begin{bmatrix} a_{1,1} & a_{1,2} & a_{1,3} & a_{1,4} \\ a_{2,1} & a_{2,2} & a_{2,3} & a_{2,4} \\ a_{3,1} & a_{3,2} & a_{3,3} & a_{3,4} \\ a_{4,1} & a_{4,2} & a_{4,3} & a_{4,4} \end{bmatrix} \begin{bmatrix} x_1 \\ x_2 \\ x_3 \\ x_4 \end{bmatrix} \tag{5.57}$$

in the form of the sum

$$\begin{bmatrix} \dot{x}_1 \\ \dot{x}_2 \end{bmatrix} = \begin{bmatrix} a_{1,1} & a_{1,2} \\ a_{2,1} & a_{2,2} \end{bmatrix} \begin{bmatrix} x_1 \\ x_2 \end{bmatrix} + \begin{bmatrix} a_{1,3} & a_{1,4} \\ a_{2,3} & a_{2,4} \end{bmatrix} \begin{bmatrix} x_3 \\ x_4 \end{bmatrix} \tag{5.58}$$

which after taking (5.56) into consideration assumes the form

$$\begin{bmatrix} \dot{x}_1 \\ \dot{x}_2 \end{bmatrix} = \mathbf{A}_r \begin{bmatrix} x_1 \\ x_2 \end{bmatrix} \tag{5.59}$$

where matrix $\mathbf{A}_r$ being sought is

$$\mathbf{A}_r = \begin{bmatrix} a_{1,1} & a_{1,2} \\ a_{2,1} & a_{2,2} \end{bmatrix} + \begin{bmatrix} a_{1,3} & a_{1,4} \\ a_{2,3} & a_{2,4} \end{bmatrix} \begin{bmatrix} x_{3,1} & x_{3,2} \\ x_{4,1} & x_{4,2} \end{bmatrix} \begin{bmatrix} x_{1,1} & x_{1,2} \\ x_{2,1} & x_{2,2} \end{bmatrix}^{-1}. \tag{5.60}$$

Formula (5.60) can be easily generalised for a matrix $\mathbf{A}$ with dimension $n \times n$.

$$\mathbf{A} = \begin{bmatrix} a_{1,1} & a_{1,2} & . & . & a_{1,n} \\ a_{2,1} & a_{2,2} & . & . & a_{2,n} \\ . & & . & . & . \\ . & & . & . & . \\ a_{n,1} & a_{n,2} & . & . & a_{n,n} \end{bmatrix} \tag{5.61}$$

where in the reduced matrix $\mathbf{A}_{r(l\times l)}$ only l eigenvalues $\lambda_r, \lambda_s, \dots \lambda_t$ for $n > l$ remain. It can be easily shown that matrix $\mathbf{A}_r$ is then [15]:

$$\mathbf{A}_r = \begin{bmatrix} a_{r,r} & a_{r,s} & a_{r,t} & . & . \\ a_{s,r} & a_{s,s} & a_{s,t} & . & . \\ a_{t,r} & a_{t,s} & a_{t,t} & . & . \\ . & . & . & . & . \\ . & . & . & . & . \end{bmatrix}$$

$$+ \begin{bmatrix} a_{r,1} & . & a_{r,r-1} & a_{r,r+1} & . & a_{r,s-1} & a_{r,s+1} & . & a_{r,n} \\ a_{s,1} & . & a_{s,r-1} & a_{s,r+1} & . & a_{s,s-1} & a_{s,s+1} & . & a_{s,n} \\ a_{t,1} & . & a_{t,r-1} & a_{t,r+1} & . & a_{t,s-1} & a_{t,s+1} & . & a_{t,n} \\ . & . & . & . & . & . & . & . & . \\ . & . & . & . & . & . & . & . & . \end{bmatrix}$$

$$\cdot \begin{bmatrix} x_{1,1} & x_{1,2} & . & . & x_{1,l} \\ . & . & . & . & . \\ x_{r-1,1} & x_{r-1,2} & . & . & x_{r-1,l} \\ x_{r+1,1} & x_{r+1,2} & . & . & x_{r+1,l} \\ . & . & . & . & . \\ x_{s-1,1} & x_{s-1,2} & . & . & x_{s-1,l} \\ x_{s+1,1} & x_{s+1,2} & . & . & x_{s+1,l} \\ . & . & . & . & . \\ x_{t-1,1} & x_{t-1,2} & . & . & x_{t-1,l} \\ x_{t+1,1} & x_{t+1,2} & . & . & x_{t+1,l} \\ . & . & . & . & . \\ x_{n,1} & x_{n,2} & . & . & x_{n,l} \end{bmatrix} \begin{bmatrix} x_{r,1} & x_{r,2} & . & . & x_{r,l} \\ x_{s,1} & x_{s,2} & . & . & x_{s,l} \\ x_{t,1} & x_{t,2} & . & . & x_{t,l} \\ . & . & . & . & . \\ . & . & . & . & . \end{bmatrix}^{-1} \cdot \tag{5.62}$$

5.5. Simplification of models using the Routh table coefficients

The model simplification method using the Routh table coefficients is simple in use and gives good results in a series of cases of reduction of the high orders of models [35]. To this end, it makes use of two consecutive lines of Routh table coefficients, determined separately for the polynomials of the numerator and denominator of the primary model transfer function. For model (2.24) given in the form of a quotient of two polynomials, we can determine the Routh table for the numerator in the form

$$
\begin{array}{cccccc}
b_{1,1} & b_{1,2} & b_{1,3} & b_{1,4} & \cdots \\
b_{2,1} & b_{2,2} & b_{2,3} & b_{2,4} & \cdots \\
b_{3,1} & b_{3,2} & b_{3,3} \\
b_{4,1} & b_{4,2} & b_{4,3} \\
\cdots\cdots\cdots \\
b_{m,1} \\
b_{m+1,1}
\end{array}
\tag{5.63}
$$

and in a similar form for the denominator

$$
\begin{array}{cccccc}
a_{1,1} & a_{1,2} & a_{1,3} & a_{1,4} & \cdots \\
a_{2,1} & a_{2,2} & a_{2,3} & a_{2,4} & \cdots \\
a_{3,1} & a_{3,2} & a_{3,3} \\
a_{4,1} & a_{4,2} & a_{4,3} \\
\cdots\cdots\cdots \\
a_{n,1} \\
a_{n+1,1}
\end{array}
\tag{5.64}
$$

where in (5.63)

$$
\begin{array}{cccc}
b_{1,1} = b_m & b_{1,2} = b_{m-2} & b_{1,3} = b_{m-4} & b_{1,4} = b_{m-6} \\
b_{2,1} = b_{m-1} & b_{2,2} = b_{m-3} & b_{2,3} = b_{m-5} & b_{2,4} = b_{m-7}
\end{array}
$$

$$
b_{i,j} = -\frac{1}{b_{i-1,1}} \begin{vmatrix} b_{i-2,1} & b_{i-2,j+1} \\ b_{i-1,1} & b_{i-1,j+1} \end{vmatrix} \qquad i = 3,4,\ldots n \qquad j = 1,2,\ldots
\tag{5.65}
$$

and in (5.64)

$$a_{1,1} = a_n = 1 \qquad a_{1,2} = a_{n-2} \qquad a_{1,3} = a_{n-4} \qquad a_{1,4} = a_{n-6}$$

$$a_{2,1} = a_{n-1} \qquad a_{2,2} = a_{n-3} \qquad a_{2,3} = a_{n-5} \qquad a_{2,4} = a_{n-7}$$

$$a_{i,j} = -\frac{1}{a_{i-1,1}} \begin{vmatrix} a_{i-2,1} & a_{i-2,j+1} \\ a_{i-1,1} & a_{i-1,j+1} \end{vmatrix} \qquad i = 3, 4, \ldots n \qquad j = 1, 2, \ldots . \tag{5.66}$$

It can be easily seen that the primary model (2.24) of the order n explicitly determines the two initial lines given by relationships (5.63) and (5.64), while the following lines (2) and (3), (3) and (4), (4) and (5), ... etc. permit simplified models of the order $(n\text{-}1)$, $(n\text{-}2)$, $(n\text{-}3)$, ... etc. to be determined. In this way the simplified model of the order $(n\text{-}1)$ is given by the formula

$$K_{n-1}(s) = \frac{b_{m-1}s^{m-1} + b_{3,1}s^{m-2} + b_{m-3}s^{m-3} + b_{3,2}s^{m-4} + \ldots}{a_{n-1}s^{n-1} + a_{3,1}s^{n-2} + a_{n-3}s^{n-3} + a_{3,2}s^{n-4} + \ldots} \tag{5.67}$$

and the model of the order $(n\text{-}2)$ by

$$K_{n-2}(s) = \frac{b_{3,1}s^{m-2} + b_{4,1}s^{m-3} + b_{3,2}s^{m-4} + b_{4,2}s^{m-5} + \ldots}{a_{3,1}s^{n-2} + a_{4,1}s^{n-3} + a_{3,2}s^{n-4} + a_{4,2}s^{n-5} + \ldots} . \tag{5.68}$$

Further models of lower orders can be determined in an identical way.

5.6. Simplification of models by means of Routh table and Schwarz matrix

Based on the Routh table (5.64) of the primary model and on the Schwarz canonical matrix determined from it, the primary model can be simplified. Its order can be determined in such a way that state variables eliminated from it would have a negligible effect on the system response. Let us assume that the primary model

$$\dot{\mathbf{x}}(t) = \mathbf{A}\mathbf{x}(t) + \mathbf{B}u(t) \tag{5.69}$$

is given in the phase-variable canonical form so that matrix $\mathbf{A}$ is given in the form

$$
\mathbf{A} = \begin{bmatrix}
0 & 1 & 0 & . & 0 \\
0 & 0 & 1 & . & 0 \\
. & . & . & . & . \\
 & & & . & \\
-a_0 & -a_1 & . & . & -a_{n-1}
\end{bmatrix}
\tag{5.70}
$$

then in order to determine the Schwarz matrix corresponding to it, we use the Routh table (5.64) and determine on this base the following transform matrix $\mathbf{P}$ [12]

$$
\mathbf{P} = \begin{bmatrix}
1 & . & 0 & 0 & 0 & 0 & 0 & 0 & 0 \\
. & . & . & . & . & . & . & . & . \\
\dfrac{a_{n-1,2}}{a_{n-1,1}} & . & 1 & 0 & 0 & 0 & 0 & 0 & 0 \\
0 & . & 0 & 1 & 0 & 0 & 0 & 0 & 0 \\
\dfrac{a_{n-3,3}}{a_{n-3,1}} & . & \dfrac{a_{6,2}}{a_{6,1}} & 0 & 1 & 0 & 0 & 0 & 0 \\
0 & . & 0 & \dfrac{a_{5,2}}{a_{5,1}} & 0 & 1 & 0 & 0 & 0 \\
\dfrac{a_{n-5,4}}{a_{n-5,1}} & . & \dfrac{a_{4,3}}{a_{4,1}} & 0 & \dfrac{a_{4,2}}{a_{4,1}} & 0 & 1 & 0 & 0 \\
0 & . & 0 & \dfrac{a_{3,3}}{a_{3,1}} & 0 & \dfrac{a_{3,2}}{a_{3,1}} & 0 & 1 & 0 \\
. & . & \dfrac{a_{2,4}}{a_{2,1}} & 0 & \dfrac{a_{2,3}}{a_{2,1}} & 0 & \dfrac{a_{2,2}}{a_{2,1}} & 0 & 1
\end{bmatrix}
\tag{5.71}
$$

In this matrix each element of the main diagonal is unity, each element of the first lower and the following odd diagonals is zero, while elements of the second and of the following even diagonals are determined on the base of the Routh table. Using matrices $\mathbf{A}$, $\mathbf{P}$ and the inverse matrix $\mathbf{P}^{-1}$ we can determine the Schwarz matrix $\mathbf{G}$ from the following product

$$
\mathbf{G} = \mathbf{PAP}^{-1} = \begin{bmatrix}
0 & 1 & 0 & . & 0 & 0 \\
-g_1 & 0 & 1 & . & 0 & 0 \\
0 & -g_2 & 0 & . & 0 & 0 \\
. & . & . & . & . & . \\
0 & 0 & 0 & . & 0 & 1 \\
0 & 0 & 0 & . & -g_{n-1} & -g_n
\end{bmatrix}
\tag{5.72}
$$

and depending on the dimension of matrix $\mathbf{A}_{(nxn)}$, the matrices $\mathbf{P}_{(2x2)}$, $\mathbf{P}_{(3x3)}$, $\mathbf{P}_{(4x4)}$, ... etc. are determined by means of the square matrices of the respective dimensions of the right-hand side lower corner matrix $\mathbf{P}$ (5.71). For $n = 2, 3$ and 4 the respective matrices $\mathbf{P}_{(2x2)}$, $\mathbf{P}_{(3x3)}$, $\mathbf{P}_{(4x4)}$ assume the simple form

$$\mathbf{P}_{(2x2)} = \begin{bmatrix} 1 & 0 \\ 0 & 1 \end{bmatrix} \tag{5.73}$$

$$\mathbf{P}_{(3x3)} = \begin{bmatrix} 1 & 0 & 0 \\ 0 & 1 & 0 \\ \dfrac{a_{2,2}}{a_{2,1}} & 0 & 1 \end{bmatrix} \tag{5.74}$$

$$\mathbf{P}_{(4x4)} = \begin{bmatrix} 1 & 0 & 0 & 0 \\ 0 & 1 & 0 & 0 \\ \dfrac{a_{3,2}}{a_{3,1}} & 0 & 1 & 0 \\ 0 & \dfrac{a_{2,2}}{a_{2,1}} & 0 & 1 \end{bmatrix} \tag{5.75}$$

....

etc.

If the primary model (5.69) is given in arbitrary form and the system is controllable, i.e. condition (2.38) is satisfied, then the model can be simplified to the Schwarz canonical form

$$\dot{\mathbf{z}}(t) = \mathbf{G}\mathbf{z}(t) + \mathbf{F}u(t) \tag{5.76}$$

by means of transformation

$$\mathbf{x}(t) = \mathbf{H}\mathbf{z}(t) \tag{5.77}$$

where in (5.76) matrix $\mathbf{G}$ is the Schwarz matrix and

$$\mathbf{F} = \begin{bmatrix} 0 \\ 0 \\ . \\ . \\ 1 \end{bmatrix}. \tag{5.78}$$

In order to determine matrix $\mathbf{H}$ we proceed identically as in the case of (2.42) - (2.51) inserting in place of matrix $\mathbf{A}_0$ (2.41) the Schwarz matrix $\mathbf{G}$ (5.72). Then from (2.45) we have

$$\mathbf{AH} = \mathbf{HG} \tag{5.79}$$

and from (2.46)

$$\mathbf{B} = \mathbf{HF} . \tag{5.80}$$

Using formulae (2.47) - (2.51) for equations (5.79) and (5.80) we obtain the transformation matrix $\mathbf{H}$, whose consecutive columns are

$$\begin{aligned}
\mathbf{h}_n &= \mathbf{B} \\
\mathbf{h}_{n-1} &= \mathbf{A}\mathbf{h}_n + g_n \mathbf{h}_n \\
\mathbf{h}_{n-2} &= \mathbf{A}\mathbf{h}_{n-1} + g_{n-1} \mathbf{h}_n \\
&\cdots \cdots \\
\mathbf{h}_{n-k} &= \mathbf{A}\mathbf{h}_{n-k+1} + g_{n-k+1} \mathbf{h}_{n-k+2} \quad \textit{for } k = 2, 3, \dots n-1 .
\end{aligned} \tag{5.81}$$

In this way matrix $\mathbf{H}$ is determined completely if g_i are known. The coefficients g_i result from matrix (5.72) or can be determined from the first column of Routh table (5.64), resulting from the characteristic equation of matrix $\mathbf{A}$ from the following relations [5]:

$$\begin{aligned}
g_i &= \frac{a_{n-i+2,1}}{a_{n-i,1}} \quad i = 1, 2, \dots n-1 \\
g_n &= \frac{a_{2,1}}{a_{1,1}} .
\end{aligned} \tag{5.82}$$

Possibilities of reducing matrix $\mathbf{A}$ depend on the results of comparisons between elements g_i occurring in matrix $\mathbf{G}$, starting with the quotient $\dfrac{g_{n-1}}{g_{n-2}}$ and finishing at $\dfrac{g_2}{g_1}$. Thus the three following cases can occur:

- the ratio of the compared coefficients is higher than or equal to 10

$$\frac{g_{n-i}}{g_{n-i-1}} \geq 10 \qquad i = 1, 2, \ldots n-2 . \tag{5.83}$$

The multiplicity of appearance of this condition indicates by how many orders the model can be simplified,
- the ratio of compared coefficients is between 1 and 10

$$1 < \frac{g_{n-i}}{g_{n-i-1}} < 10 \tag{5.84}$$

then the eigenvalues of the matrices are of the same order hence no simplification is possible,
- if for any i the ratio of the coefficients being compared reaches a value lower than 1

$$\frac{g_{n-i}}{g_{n-i-1}} < 1 \tag{5.85}$$

then the checking of simplification possibilities is finished at that i . Otherwise it is carried on until $\dfrac{g_2}{g_1}$ is compared and if the i-th comparison is the last successful one, the system model can be simplified to the order of $m = n - i$.

In order to determine the reduced matrix $\mathbf{A}_r$, containing only those eigenvalues, which have an essential effect on the system response, we partition the equation

$$\dot{\mathbf{z}}(t) = \mathbf{G}\mathbf{z}(t) \tag{5.86}$$

into m and i components [5]:

$$\begin{bmatrix} \dot{\mathbf{z}}_m(t) \\ \dot{\mathbf{z}}_i(t) \end{bmatrix} = \begin{bmatrix} \mathbf{G}_{mm} & \mathbf{G}_{mi} \\ \mathbf{G}_{im} & \mathbf{G}_{ii} \end{bmatrix} \begin{bmatrix} \mathbf{z}_m(t) \\ \mathbf{z}_i(t) \end{bmatrix}. \tag{5.87}$$

Solution of the equation (5.87) with respect to $\dot{\mathbf{z}}_m(t)$ and $\mathbf{z}_i(t)$ with $\dot{\mathbf{z}}_i(t) \approx 0$ taken into consideration, which results from the assumption that i state variables give negligible effect, yields

$$\dot{\mathbf{z}}_m(t) = \mathbf{G}_{mm}\mathbf{z}_m(t) + \mathbf{G}_{mi}\mathbf{z}_i(t) \tag{5.88}$$

and

$$\mathbf{z}_i(t) = -\mathbf{G}_{ii}^{-1}\mathbf{G}_{im}\mathbf{z}_m(t). \tag{5.89}$$

Inserting (5.89) into (5.88) gives the simplified model in the Schwarz form

$$\dot{\mathbf{z}}_m(t) = (\mathbf{G}_{mm} - \mathbf{G}_{mi}\mathbf{G}_{ii}^{-1}\mathbf{G}_{im})\mathbf{z}_m(t) = \mathbf{G}_m\mathbf{z}_m(t). \tag{5.90}$$

Repeating similar procedures of partitioning into m and i components with respect to equation (5.77) we obtain

$$\begin{bmatrix} \mathbf{x}_m(t) \\ \mathbf{x}_i(t) \end{bmatrix} = \begin{bmatrix} \mathbf{H}_{mm} & \mathbf{H}_{mi} \\ \mathbf{H}_{im} & \mathbf{H}_{ii} \end{bmatrix} \begin{bmatrix} \mathbf{z}_m(t) \\ \mathbf{z}_i(t) \end{bmatrix} \tag{5.91}$$

which after solving with respect to $\mathbf{x}_m(t)$ and considering (5.89) gives

$$\mathbf{x}_m(t) = (\mathbf{H}_{mm} - \mathbf{H}_{mi}\mathbf{G}_{ii}^{-1}\mathbf{G}_{im})\mathbf{z}_m(t) = \mathbf{H}_m\mathbf{z}_m. \tag{5.92}$$

From (5.92) and (5.90) it follows easily that

$$\dot{\mathbf{x}}_m(t) = \mathbf{H}_m\mathbf{G}_m\mathbf{H}_m^{-1}\mathbf{x}_m(t) \tag{5.93}$$

thus the form of the reduced matrix $\mathbf{A}_r$ represents the relationship

$$\mathbf{A}_r = \mathbf{H}_m \mathbf{G}_m \mathbf{H}_m^{-1} \qquad (5.94)$$

where $\mathbf{H}_m$ and $\mathbf{G}_m$ are determined by (5.92) and (5.90)

5.7. Simplification of models by comparison of characteristic equation coefficients

The method of comparison of coefficients at the highest power s of the transfer function model's denominator permits us to neglect those poles with negative real parts, whose values $|\operatorname{Re} s_i|$ are relatively high as compared with the remaining ones. The principle of neglecting poles was presented in [86]. It is stated that the i-th pole can be neglected if its real part is r-times higher (e.g. $r = 10$) than the smallest real part $|\operatorname{Re} s_i|_{\min}$ of the pole appearing in the model

$$|\operatorname{Re} s_i| \geq r |\operatorname{Re} s_i|_{\min} . \qquad (5.95)$$

At the same time a correction factor is introduced into the numerator of the model simplified in this manner, which ensures equality of the static amplifications.

Let us consider the general case of a high order model with multiple poles of the order n and m single poles with values significantly higher. The denominator of such a model can be written as follows

$$M(s) = (1 + sT)^n \prod_{i=1}^{m} (1 + sTa_i) \qquad (5.96)$$

and it can be transformed to the form of a sum

$$M(s) = s^{n+m} T^{n+m} \prod_{i=1}^{m} (a_i) + s^{n+m-1} T^{n+m-1} \left(n + \sum_{i=1}^{m} \frac{1}{a_i} \right) \prod_{i=1}^{m} (a_i)$$

$$+ \ldots + sT \left(n + \sum_{i=1}^{m} \frac{1}{a_i} \right) + 1. \qquad (5.97)$$

If we insert

$$s = \frac{S}{T} \tag{5.98}$$

into formula (5.97) then $M(s)$ assumes the form

$$M(s) = S^{n+m} \prod_{i=1}^{m} (a_i) + S^{n+m-1} \left(n + \sum_{i=1}^{m} \frac{1}{a_i} \right) \prod_{i=1}^{m} (a_i) + ... + S \left(n + \sum_{i=1}^{m} \frac{1}{a_i} \right) + 1. \tag{5.99}$$

In [86] it was shown that if in (5.99) the coefficient at the highest power of S is significantly lower than 1, it can be neglected and the rejection of the expression $S^{n+m} \prod_{i=1}^{m} (a_i)$ causes only a small change in the values of poles of the model (5.96).

Thus we begin to simplify a model with its denominator given by equation (5.99) by assuming the order of the simplified model is $k = n + m - 1$, substituting s given by formula

$$s = \frac{kS}{T \left(n + \sum_{i=1}^{m} \frac{1}{a_i} \right)} \tag{5.100}$$

and checking if the coefficient at S^{n+m} is significantly lower then 1. If so, we reduce the model by 1, assume a new $k = n + m - 2$ and again check the value of the coefficient at the highest power of S. The simplification procedure can be repeated until the coefficient at the highest power of S becomes significantly lower than 1.

5.8. Examples

Example 5.1

A model of a system is given in the form of the polynomial $f(t) = t^3 + t^5$. Using the Chebyshev polynomials determine the simplified model in the form of a polynomial of the degree $n = 3$, over the interval [0, 2].

Solution

For the Chebyshev polynomial (5.29) we compute the coefficients a_0 and a_k using formula (5.28), where $w(t)$ is determined by formula (5.26) and $t_f = 2$. Hence we have

$$a_0 = \frac{1}{\pi} \int\limits_0^2 \frac{t^3 + t^5}{\sqrt{2t - t^2}} (1)\,dt = 10.375$$

$$a_1 = \frac{2}{\pi} \int\limits_0^2 \frac{t^3 + t^5}{\sqrt{2t - t^2}} \left(1 - 2\left(\frac{t}{2}\right)\right) dt = -16.875$$

$$a_2 = \frac{2}{\pi} \int\limits_0^2 \frac{t^3 + t^5}{\sqrt{2t - t^2}} \left(1 - 8\left(\frac{t}{2}\right) + 8\left(\frac{t}{2}\right)^2\right) dt = 9$$

$$a_3 = \frac{2}{\pi} \int\limits_0^2 \frac{t^3 + t^5}{\sqrt{2t - t^2}} \left(1 - 18\left(\frac{t}{2}\right) + 48\left(\frac{t}{2}\right)^2 - 32\left(\frac{t}{2}\right)^3\right) dt = -3.062.$$

(5.101)

The polynomial being sought has the form

$$T_3(t) = \sum_{k=0}^{3} a_k T_k(t) = 10.375 - 16.875\left(1 - 2\frac{t}{2}\right)$$

$$+ 9\left(1 - 8\frac{t}{2} + 8\left(\frac{t}{2}\right)^2\right) - 3.062\left(1 - 18\frac{t}{2} + 48\left(\frac{t}{2}\right)^2 - 32\left(\frac{t}{2}\right)^3\right)$$

(5.102)

and after reductions

$$T_3(t) = -0.562 + 8.433t - 18.744t^2 + 12.248t^3.$$

(5.103)

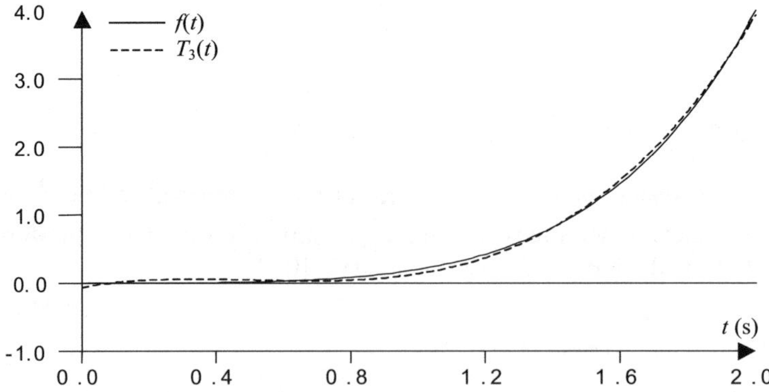

Fig. 5.1. Approximation of $f(t) = t^3 + t^5$ by means of polynomial $T_3(t)$ (5.103).

Example 5.2

Simplify the polynomial $f(t)$ from the previous example making use of Legendre orthonormal polynomials of the degree $n = 3$, satisfying the condition given by formula (5.1) over the interval $[-1, 1]$.

Solution

In order to obtain the Legendre orthonormal polynomials $\Phi(t)$ let us divide polynomials (5.30) - (5.33) by the square roots of their modules. We have thus

$$\Phi_0(t) = \frac{1}{\sqrt{\int\limits_{-1}^{1} dt}} = \frac{\sqrt{2}}{2}$$

$$\Phi_1(t) = \frac{t}{\sqrt{\int\limits_{-1}^{1} t^2\, dt}} = \frac{\sqrt{6}}{2} t$$

$$\Phi_2(t) = \frac{t^2 - \dfrac{1}{3}}{\sqrt{\int\limits_{-1}^{1} (t^2 - \dfrac{1}{3})(t^2 - \dfrac{1}{3})\, dt}} = \frac{\sqrt{10}}{4}(5t^3 - 3t)$$

$$\Phi_3(t) = \frac{t^3 - \dfrac{3}{5}t}{\sqrt{\int\limits_{-1}^{1} (t^3 - \dfrac{3}{5}t)(t^2 - \dfrac{3}{5}t)\, dt}} = \frac{\sqrt{14}}{4}(5t^3 - 3t).$$

$$(5.104)$$

We determine the coefficients a_k of the polynomial being sought on the basis of formula (5.6) for which $w(t) = 1$ and $\alpha_k = 1$

$$a_1 = \int\limits_{-1}^{1} (t^3 + t^5)\frac{\sqrt{2}}{2}\, dt = 0$$

$$a_1 = \int\limits_{-1}^{1} (t^3 + t^5)\frac{\sqrt{6}}{4}\, t\, dt = \frac{12\sqrt{6}}{35}$$

$$a_2 = \int\limits_{-1}^{1} (t^3 + t^5)\frac{\sqrt{10}}{4}(5t^2 - 1)\, dt = 0$$

$$a_3 = \int\limits_{-1}^{1} (t^3 + t^5)\frac{\sqrt{14}}{4}(5t^3 - 3t)\, dt = \frac{38\sqrt{14}}{315}.$$

$$(5.105)$$

The polynomial being sought is thus

$$\Phi_3(t) = \sum_{k=0}^{3} a_k \Phi_k(t) = \frac{12\sqrt{6}}{35}\frac{\sqrt{6}}{2}t + \frac{38\sqrt{14}}{315}\frac{\sqrt{14}}{4}(5t^3 - 3t) \qquad (5.106)$$

from whence we obtain finally after reductions

$$\Phi_3(t) = \frac{-5}{21}t + \frac{19}{9}t^3 . \qquad (5.107)$$

The quality of approximation is shown in Fig. 5.2.

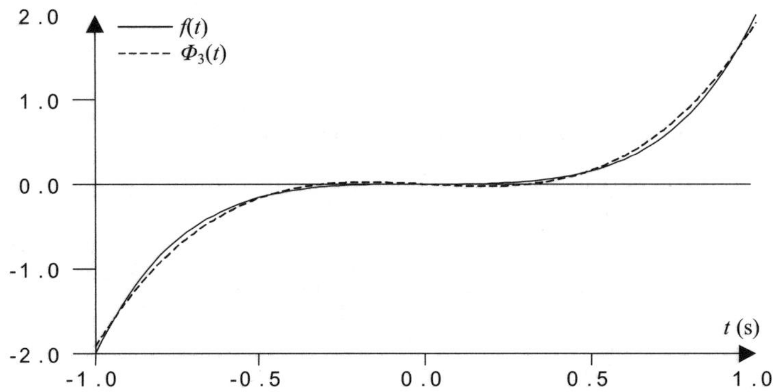

Fig. 5.2. Approximation of $f(t) = t^3 + t^5$ by means of polynomial $\Phi_3(t)$ (5.107) .

Example 5.3

Simplify the model of the third order

$$K_3(s) = \frac{s+5}{s^3 + 8s^2 + 19s + 12} \qquad (5.108)$$

to a model of the second order, given in the form

$$K_2(s) = \frac{d_1 s + d_0}{s^2 + c_1 s + c_0} \tag{5.109}$$

and satisfying condition (5.36). Assume the lower integration limit is $\omega_{\min} = \omega_1 = 0$ while the upper integration limit $\omega_{\max} = \omega_2$ should correspond to the ordinate for which the frequency response does not exceed 3% of the value for $\omega = 0$.

Solution

For the spectral transmittances of models

$$K_2(j\omega) = \frac{j\omega d_1 + d_0}{(c_0 - \omega^2) + j\omega c_1} \tag{5.110}$$

$$K_3(j\omega) = \frac{j\omega + 5}{(12 - 8\omega^2) + j\omega(19 - \omega^2)} \tag{5.111}$$

the error $I_2(\omega)$ is

$$I_2(\omega) = \int_{\omega_1}^{\omega_2} \left\{ \left[(5c_0 - 12d_0) - (5 + c_1 - 8d_0 - 19d_1)\omega^2 - d_1\omega^4 \right]^2 \right. \tag{5.112}$$
$$\left. + \left[(5c_1 + c_0 - 19d_0 - 12d_1)\omega - (1 - d_0 - 8d_1) \right]^2 \right\} d\omega.$$

Having integrated the above equation, computed the derivatives $\dfrac{dI_2(\omega)}{dc_0}$, $\dfrac{dI_2(\omega)}{dc_1}$, $\dfrac{dI_2(\omega)}{dd_0}$ and $\dfrac{dI_2(\omega)}{dd_1}$ and having equated them to zero and after substituting $\omega_1 = 0$ we obtain the following equations

$$\left(\frac{166}{3}c_0 - 40c_1 + 96d_1 \right)\omega_2^3 + \left(\frac{338}{5}d_1 + \frac{6}{5}c_0 - \frac{166}{5} + \frac{42}{5}c_1 \right)\omega_2^5 \tag{5.113}$$
$$+ \left(\frac{-6}{7} + \frac{52}{7}d_1 + \frac{2}{7}c_1 \right)\omega_2^7 + \frac{2}{9}d_1\omega_2^9 = 0$$

$$\left(288d_0 - 120c_0\right)\omega_2 + \left(14c_0 + 40 - \frac{166}{3}c_1 + \frac{338}{3}d_0\right)\omega_2^3$$
$$+ \left(\frac{52}{5}d_0 - \frac{42}{5} + \frac{2}{5}c_0 - \frac{6}{5}c_1\right)\omega_2^5 = 0 \tag{5.114}$$

$$\left(\frac{-166}{3}d_0 - 40d_1 + \frac{50}{3}c_1\right)\omega_2^3 + \left(\frac{42}{5}d_1 + \frac{2}{5}c_1 - \frac{6}{5}d_0\right)\omega_2^5 + \frac{2}{7}d_1\omega_2^7 = 0 \tag{5.115}$$

$$\left(50c_0 - 120d_0\right)\omega_2 + \left(14d_0 - \frac{50}{3} + \frac{166}{3}d_1 + \frac{2}{3}c_0\right)\omega_2^3$$
$$+ \left(\frac{-2}{5} + \frac{2}{5}d_0 + \frac{6}{5}d_1\right)\omega_2^5 = 0. \tag{5.116}$$

Let us assume the upper integration limit $\omega_2 = 10$ for which the ordinate of the frequency response is $9.91 \cdot 10^{-3}$, which is 2.37% of its value for $\omega = 0$. The solution to the equations for $\omega_2 = 10$ gives: $d_0 = 1.015$, $d_1 = -1.405 \cdot 10^{-3}$, $c_0 = 2.826$, $c_1 = 3.233$, thus the simplified model $K_2(s)$ has the form

$$K_2(s) = \frac{-1.405 \cdot 10^{-3}s + 1.015}{s^2 + 3.233s + 2.826}. \tag{5.117}$$

The diagrams in Fig. 5.3 and Fig. 5.4 show frequency responses $\left|K_3(j\omega)\right|$ and $\left|K_2(j\omega)\right|$ as well as impulse responses $k_3(t)$ and $k_2(t)$ of the models $K_3(s)$ and $K_2(s)$.

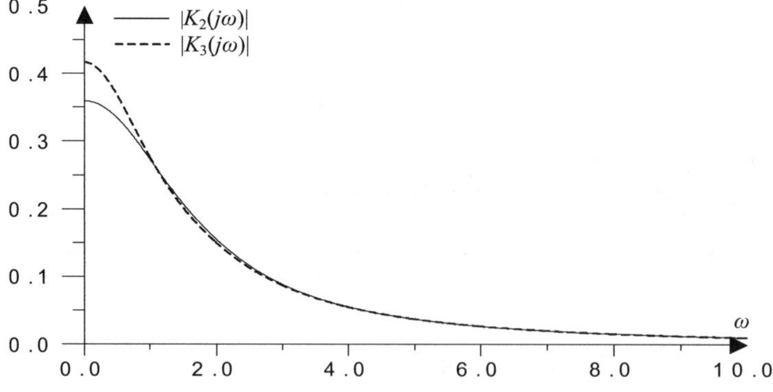

Fig. 5.3. Frequency responses $\left|K_3(j\omega)\right|$ and $\left|K_2(j\omega)\right|$ of models $K_3(s)$ (5.108) and $K_2(s)$ (5.117).

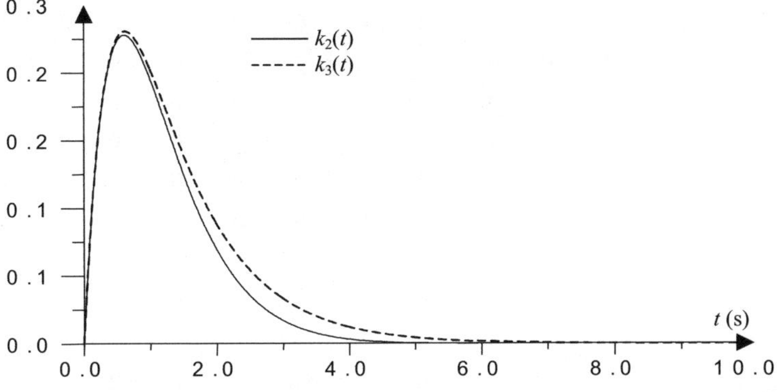

Fig. 5.4. Impulse responses $k_3(t)$ and $k_2(t)$ of models $K_3(s)$ (5.108) and $K_2(s)$ (5.117) .

Example 5.4

The sixth order model is given in the form

$$K_6(s) = \frac{6s^4 + 50s^3 + 196s^2 + 418s + 434}{s^6 + 12s^5 + 71s^4 + 256s^3 + 575s^2 + 804s + 585}. \tag{5.118}$$

Simplify this model to the order $q = 4$ in such a way so that the impulse response of the simplified model $K_4(s)$ expanded into a power series, will have the first eight terms equal with the first eight terms of the impulse response of the model $K_6(s)$.

Solution

The values of the parameters of the model $K_6(s)$ are as follows: $b_{65} = 0$, $b_{64} = 6$, $b_{63} = 50$, $b_{62} = 196$, $b_{61} = 418$, $b_{60} = 434$, $a_{65} = 12$, $a_{64} = 71$, $a_{63} = 256$, $a_{62} = 575$, $a_{61} = 804$, $a_{60} = 585$. For a model of the order $n = 6$ formula (4.50) assumes the form of the system equations (5.119). After substituting parameters of the model $K_6(s)$ into (5.119) and solving it we obtain: $A_{60} = 0$, $A_{61} = 6$, $A_{62} = -22$,

$A_{63} = 34$, $\quad A_{64} = 36$, $\quad A_{65} = -230$, $\quad A_{66} = -674$, $\quad A_{67} = 9.83 \cdot 10^3$,

$A_{68} = -4.639 \cdot 10^4$, $A_{69} = 1.147 \cdot 10^5$, $A_{610} = -4.805 \cdot 10^4$, $A_{611} = -6.69 \cdot 10^5$.

$$
\begin{bmatrix} A_{60} \\ A_{61} \\ A_{62} \\ A_{63} \\ A_{64} \\ A_{65} \\ A_{66} \\ A_{67} \\ A_{68} \\ A_{69} \\ A_{610} \\ A_{611} \end{bmatrix} =
\begin{bmatrix}
1 & 0 & 0 & 0 & 0 & 0 & 0 & 0 & 0 & 0 & 0 & 0 \\
a_{65} & 1 & 0 & 0 & 0 & 0 & 0 & 0 & 0 & 0 & 0 & 0 \\
a_{64} & a_{65} & 1 & 0 & 0 & 0 & 0 & 0 & 0 & 0 & 0 & 0 \\
a_{63} & a_{64} & a_{65} & 1 & 0 & 0 & 0 & 0 & 0 & 0 & 0 & 0 \\
a_{62} & a_{63} & a_{64} & a_{65} & 1 & 0 & 0 & 0 & 0 & 0 & 0 & 0 \\
a_{61} & a_{62} & a_{63} & a_{64} & a_{65} & 1 & 0 & 0 & 0 & 0 & 0 & 0 \\
a_{60} & a_{61} & a_{62} & a_{63} & a_{64} & a_{65} & 1 & 0 & 0 & 0 & 0 & 0 \\
0 & a_{60} & a_{61} & a_{62} & a_{63} & a_{64} & a_{65} & 1 & 0 & 0 & 0 & 0 \\
0 & 0 & a_{60} & a_{61} & a_{62} & a_{63} & a_{64} & a_{65} & 1 & 0 & 0 & 0 \\
0 & 0 & 0 & a_{60} & a_{61} & a_{62} & a_{63} & a_{64} & a_{65} & 1 & 0 & 0 \\
0 & 0 & 0 & 0 & a_{60} & a_{61} & a_{62} & a_{63} & a_{64} & a_{65} & 1 & 0 \\
0 & 0 & 0 & 0 & 0 & a_{60} & a_{61} & a_{62} & a_{63} & a_{64} & a_{65} & 1
\end{bmatrix}^{-1}
\begin{bmatrix} b_{65} \\ b_{64} \\ b_{63} \\ b_{62} \\ b_{61} \\ b_{60} \\ 0 \\ 0 \\ 0 \\ 0 \\ 0 \\ 0 \end{bmatrix}.
$$

$$(5.119)$$

We compute the parameters of the model $K_4(s)$ by inserting the values of the first eight $A_{60} \div A_{67}$ of the above determined coefficients into formula (4.42). For $q = 4$, this takes the form

$$
\begin{bmatrix} b_{43} \\ b_{42} \\ b_{41} \\ b_{40} \\ a_{43} \\ a_{42} \\ a_{41} \\ a_{40} \end{bmatrix} =
\begin{bmatrix}
1 & 0 & 0 & 0 & 0 & 0 & 0 & 0 \\
0 & 1 & 0 & 0 & -A_{40} & 0 & 0 & 0 \\
0 & 0 & 1 & 0 & -A_{41} & -A_{40} & 0 & 0 \\
0 & 0 & 0 & 1 & -A_{42} & -A_{41} & -A_{40} & 0 \\
0 & 0 & 0 & 0 & -A_{43} & -A_{42} & -A_{41} & -A_{40} \\
0 & 0 & 0 & 0 & -A_{44} & -A_{43} & -A_{42} & -A_{41} \\
0 & 0 & 0 & 0 & -A_{45} & -A_{44} & -A_{43} & -A_{42} \\
0 & 0 & 0 & 0 & -A_{46} & -A_{45} & -A_{44} & -A_{43}
\end{bmatrix}^{-1}
\begin{bmatrix} A_{40} \\ A_{41} \\ A_{42} \\ A_{43} \\ A_{44} \\ A_{45} \\ A_{46} \\ A_{47} \end{bmatrix}
$$

$$(5.120)$$

where $A_{40} = A_{60}$, $A_{41} = A_{61}$, $A_{42} = A_{62}$, $A_{43} = A_{63}$, $A_{44} = A_{64}$, $A_{45} = A_{65}$, $A_{46} = A_{66}$, $A_{47} = A_{67}$. Solution of the latter system of equations gives as a result: $b_{43} = 0$, $b_{42} = 6$, $b_{41} = 44.309$, $b_{40} = 136.775$, $a_{43} = 11.052$, $a_{42} = 57.652$, $a_{41} = 142.763$, $a_{40} = 168.798$. The model being sought is thus in the form

$$K_4(s) = \frac{6s^2 + 44.309s + 136.775}{s^4 + 11.052s^3 + 57.652s^2 + 142.763s + 168.798}. \tag{5.121}$$

It can be easily shown that in the impulse responses of the models $k_6(t)$ and $k_4(t)$ represented in the form of power series

$$k_6(t) = \sum_{k=0}^{\infty} c_{6k} t^k = 6.00t - 11.00t^2 + 5.66t^3 + 1.50t^4 - 1.91t^5$$
$$- 0.96t^6 + 1.95t^7 - 1.15t^8 + 0.32t^9 - 0.013t^{10} - ... + \tag{5.122}$$

and

$$k_4(t) = \sum_{k=0}^{\infty} c_{6k} t^k = 6.00t - 11.00t^2 + 5.70t^3 + 1.39t^4 - 1.76t^5$$
$$- 1.053t^6 + 1.98t^7 - 1.05t^8 + 0.99t^9 + 0.205t^{10} - ... + \tag{5.123}$$

the first $k = 2q = 8$ coefficients, with the accuracy of calculation, are equal. Hence the difference between models $K_6(s)$ and $K_4(s)$ is caused by the differences between the consecutive coefficients, starting from $k = 9$ at t^8

$$k_{6,4}(t) = k_6(t) - k_4(t) = (c_{6,8} - c_{4,8})t^8 + (c_{6,9} - c_{4,9})t^9 + ... \tag{5.124}$$

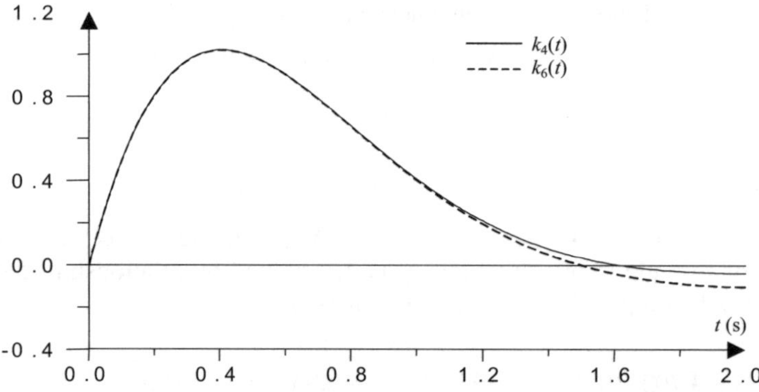

Fig. 5.5. Impulse responses of the primary model $k_6(t)$ (5.118) and the simplified model $k_4(t)$ (5.121).

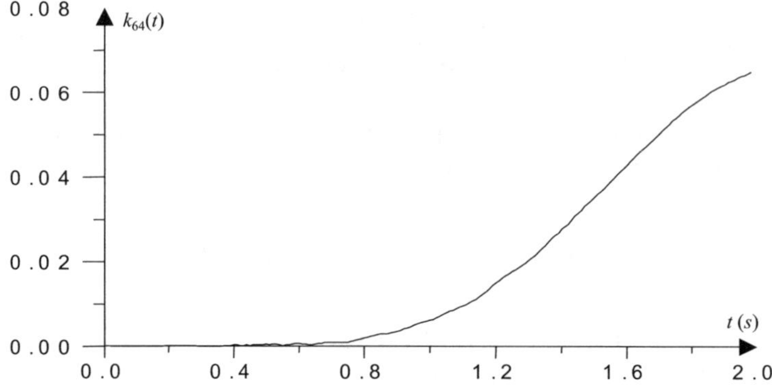

Fig. 5.6. Difference $k_{6,4}(t)$ (5.124) between the impulse responses $k_6(t)$ and $k_4(t)$.

Example 5.5

For matrix $\mathbf{A}_{(4x4)}$

$$\mathbf{A} = \begin{bmatrix} 0 & 1 & 0 & 0 \\ 0 & 0 & 1 & 0 \\ 0 & 0 & 0 & 1 \\ -7500 & -8800 & -1365 & -66 \end{bmatrix} \tag{5.125}$$

which has the following eigenvaues: $\lambda_2 = -30$, $\lambda_2 = -25$, $\lambda_3 = -10$, $\lambda_4 = -1$ determine simplified matrix $\mathbf{A}_r$ with retained eigenvalues: $\lambda_2 = -30$ and $\lambda_2 = -25$.

Solution

In the case of matrix $\mathbf{A}_{(4x4)}$ and retaining in $\mathbf{A}_{r\,(2x2)}$ the eigenvalues λ_1 and λ_2 we make use of formula (5.59) in which $\mathbf{A}_r$ is determined by relationship (5.60). The modal matrix $\mathbf{S}$ computed on the basis of matrix $\mathbf{A}$ is

$$\mathbf{S} = \begin{bmatrix} -3.702 \cdot 10^{-5} & -6.395 \cdot 10^{-5} & -9.95 \cdot 10^{-4} & -05 \\ 1.11 \cdot 10^{-3} & 1.599 \cdot 10^{-3} & 9.95 \cdot 10^{-3} & 0.5 \\ -0.033 & -0.04 & -0.099 & -0.5 \\ 0.999 & 0.999 & 0.995 & 0.5 \end{bmatrix}. \tag{5.126}$$

Thus

$$\mathbf{A}_r = \begin{bmatrix} 0 & 1 \\ 0 & 0 \end{bmatrix} + \begin{bmatrix} 0 & 0 \\ 1 & 0 \end{bmatrix} \begin{bmatrix} -0.033 & -0.04 \\ 0.999 & 0.999 \end{bmatrix}$$
$$\cdot \begin{bmatrix} -3.702 \cdot 10^{-5} & -6.395 \cdot 10^{-5} \\ 1.11 \cdot 10^{-3} & 1.599 \cdot 10^{-3} \end{bmatrix}^{-1} \tag{5.127}$$

and after calculation

$$\mathbf{A}_r = \begin{bmatrix} 0 & 1 \\ -709.698 & -53.399 \end{bmatrix}. \tag{5.128}$$

Matrix $\mathbf{A}_r$ has the following eigenvalues $\lambda_1 = -28.48$ and $\lambda_2 = -24.92$.

Example 5.6

For matrix $\mathbf{A}$ from the example 5.5 determine the simplified matrix $\mathbf{A}_r$ retaining eigenvalues $\lambda_3 = -10$ and $\lambda_4 = -1$.

Solution

In the case of retaining in $\mathbf{A}_r$ the eigenvalues λ_3 and λ_4 formula (5.59) assumes the form

$$\begin{bmatrix} \dot{x}_3 \\ \dot{x}_4 \end{bmatrix} = \mathbf{A}_r \begin{bmatrix} x_3 \\ x_4 \end{bmatrix} \tag{5.129}$$

in which $\mathbf{A}_r$ is determined by the relationship

$$\mathbf{A}_r = \begin{bmatrix} a_{33} & a_{34} \\ a_{43} & a_{44} \end{bmatrix} + \begin{bmatrix} a_{31} & a_{32} \\ a_{41} & a_{42} \end{bmatrix} \begin{bmatrix} x_{13} & x_{14} \\ x_{23} & x_{24} \end{bmatrix} \begin{bmatrix} x_{33} & x_{34} \\ x_{43} & x_{44} \end{bmatrix}^{-1}. \tag{5.130}$$

Inserting respective values into the above matrices we obtain

$$\mathbf{A}_r = \begin{bmatrix} 0 & 1 \\ -1365 & -66 \end{bmatrix} + \begin{bmatrix} 0 & 0 \\ -7500 & -8800 \end{bmatrix}$$
$$\cdot \begin{bmatrix} -9.95\cdot 10^{-4} & -0.5 \\ 9.95\cdot 10^{-3} & 0.5 \end{bmatrix} \begin{bmatrix} -0.099 & -0.5 \\ 0.995 & 0.5 \end{bmatrix}^{-1} \tag{5.131}$$

which, after computation, gives

$$\mathbf{A}_r = \begin{bmatrix} 0 & 1 \\ -10.756 & -11.756 \end{bmatrix}. \tag{5.132}$$

Matrix $\mathbf{A}_r$ has the following eigenvalues $\lambda_3 = -10.75$ and $\lambda_4 = -1.00$.

Example 5.7

Simplify the model $K_7(s)$ [72]:

$$K_7(s) = \frac{375000(s + 0.08333)}{s^7 + 83.64\,s^6 + 4097\,s^5 + 70342\,s^4 + 853703\,s^3 \cdots} \tag{5.133}$$

$$\frac{\cdots}{\cdots + 2814271 s^2 + 3310875\,s + 281250}$$

to the model $K_3(s)$ using the Routh coefficients method.

Solution

The Routh table of the model denominator is in the form

$$
\begin{array}{llll}
1 & 4097 & 853703 & 3310875 \\
83.64 & 70342 & 2814271 & 281250 \\
3.256\cdot 10^3 & 8.201\cdot 10^5 & 3.308\cdot 10^6 & \\
4.928\cdot 10^4 & 2.729\cdot 10^6 & 281250 & \\
6.398\cdot 10^5 & 3.289\cdot 10^6 & & \\
2.476\cdot 10^6 & 281250 & & \\
3.216\cdot 10^6. & & &
\end{array}
\tag{5.134}
$$

We determine the denominator of a model of the order of $n-4=3$ using formulae (5.67) and (5.68) from the 5-*th* and 6-*th* line of the Routh table. The model sought is thus in the form

$$K_3(s) = \frac{375000(s + 0.08333)}{6.398 \cdot 10^5 s^3 + 2.476 \cdot 10^6 s^2 + 3.289 \cdot 10^6 s + 281250} \tag{5.135}$$

and after reduction

$$K_3(s) = \frac{0.586s + 0.049}{s^3 + 3.87s^2 + 5.141s + 0.44}. \tag{5.136}$$

The quality of approximation obtained is shown by the impulse responses of the models $k_7(t)$ and $k_3(t)$ - Fig. 5.7

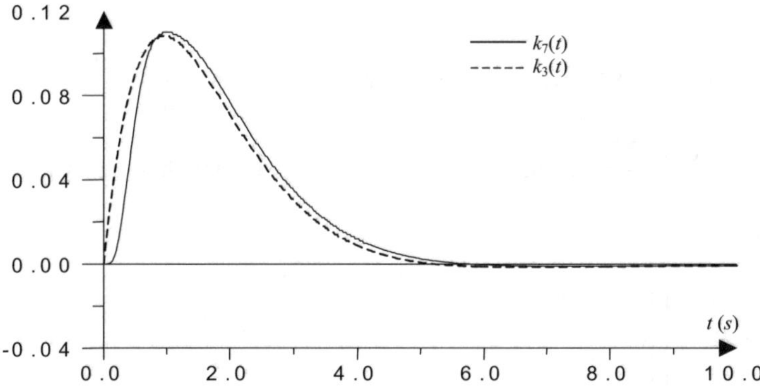

Fig. 5.7. Impulse responses of the primary model $k_7(t)$ (5.133)and the simplified model $k_3(t)$ (5.136) .

Example 5.8

The state equation from example 2.7 is given as

$$\dot{\mathbf{x}}(t) = \mathbf{A}\mathbf{x}(t) + \mathbf{B}u(t)$$

$$\mathbf{A} = \begin{bmatrix} 7 & 2 & 0 \\ 3 & 5 & -1 \\ 0 & 5 & -6 \end{bmatrix} \qquad \mathbf{B} = \begin{bmatrix} 1 \\ 1 \\ 1 \end{bmatrix} \qquad n = 3. \tag{5.137}$$

Determine the matrix $\mathbf{H}$ which transforms this equation to Schwarz canonical form.

Solution

The characteristic polynomial of matrix $\mathbf{A}$ is

$$\lambda^3 - 6\lambda^2 - 38\lambda + 139 \tag{5.138}$$

and the table of Routh coefficients corresponding to it is

$$
\begin{array}{ll}
1 & -38 \\
-6 & 139 \\
-14.833 & \\
139. &
\end{array}
\tag{5.139}
$$

Using formulae (5.82) we calculate for $n = 3$: $g_1 = \dfrac{a_{4,1}}{a_{2,1}} = \dfrac{139}{-6} = -23.167$,

$g_2 = \dfrac{a_{3,1}}{a_{1,1}} = \dfrac{-14.833}{1} = -14.833$, $g_3 = \dfrac{a_{21}}{a_{11}} = \dfrac{-6}{1} = -6$ which after insertion into

(5.81) give

$$\mathbf{h}_3 = \mathbf{B} = \begin{bmatrix} 1 \\ 1 \\ 1 \end{bmatrix}$$

$$\mathbf{h}_2 = \mathbf{A}\mathbf{h}_3 + g_3\mathbf{h}_3 = \begin{bmatrix} 7 & 2 & 0 \\ 3 & 5 & -1 \\ 0 & 5 & -6 \end{bmatrix}\begin{bmatrix} 1 \\ 1 \\ 1 \end{bmatrix} - 6\begin{bmatrix} 1 \\ 1 \\ 1 \end{bmatrix} = \begin{bmatrix} 3 \\ 1 \\ -7 \end{bmatrix} \tag{5.140}$$

$$\mathbf{h}_1 = \mathbf{A}\mathbf{h}_2 + g_2\mathbf{h}_3 = \begin{bmatrix} 7 & 2 & 0 \\ 3 & 5 & -1 \\ 0 & 5 & -6 \end{bmatrix}\begin{bmatrix} 3 \\ 1 \\ -7 \end{bmatrix} - 14.833\begin{bmatrix} 1 \\ 1 \\ 1 \end{bmatrix} = \begin{bmatrix} 8.167 \\ 6.167 \\ 32.167 \end{bmatrix}$$

and the sought matrix $\mathbf{H} = \begin{bmatrix} \mathbf{h}_1 & \mathbf{h}_2 & \mathbf{h}_3 \end{bmatrix}$ in the form

$$\mathbf{H} = \begin{bmatrix} 8.167 & 3 & 1 \\ 6.167 & 1 & 1 \\ 32.167 & -7 & 1 \end{bmatrix}. \tag{5.141}$$

It can be easily shown that the Schwarz matrix $\mathbf{G} = \mathbf{H}^{-1}\mathbf{AH}$ determined from matrix $\mathbf{H}$ is

$$
\mathbf{G} = \begin{bmatrix} 8.167 & 3 & 1 \\ 6.167 & 1 & 1 \\ 32.167 & -7 & 1 \end{bmatrix}^{-1} \begin{bmatrix} 7 & 2 & 0 \\ 3 & 5 & -1 \\ 0 & 5 & -6 \end{bmatrix}
$$
$$
\cdot \begin{bmatrix} 8.167 & 3 & 1 \\ 6.167 & 1 & 1 \\ 32.167 & -7 & 1 \end{bmatrix} = \begin{bmatrix} 0 & 1 & 0 \\ 23.167 & 0 & 1 \\ 0 & 14.833 & 6 \end{bmatrix}
$$

(5.142)

and vector $\mathbf{F}$ (5.80) is

$$
\mathbf{F} = \mathbf{H}^{-1}\mathbf{B} = \begin{bmatrix} 8.167 & 3 & 1 \\ 6.167 & 1 & 1 \\ 32.167 & -7 & 1 \end{bmatrix}^{-1} \begin{bmatrix} 1 \\ 1 \\ 1 \end{bmatrix} = \begin{bmatrix} 0 \\ 0 \\ 1 \end{bmatrix}.
$$

(5.143)

Example 5.9

The state equation from example 2.7 is given in phase-variable canonical form

$$
\dot{\mathbf{x}}(t) = \mathbf{A}\mathbf{x}(t) + \mathbf{B}u(t)
$$

$$
\mathbf{A} = \begin{bmatrix} 0 & 1 & 0 \\ 0 & 0 & 1 \\ -139 & 38 & 6 \end{bmatrix} \qquad \mathbf{B} = \begin{bmatrix} 0 \\ 0 \\ 1 \end{bmatrix} \qquad n = 3.
$$

(5.144)

Simplify it to the Schwarz canonical form using the matrix $\mathbf{P}$ (5.71).

Solution

For a given matrix $\mathbf{A}$ matrix $\mathbf{P}$ (5.71) assumes the form

$$
\mathbf{P} = \begin{bmatrix} 1 & 0 & 0 \\ 0 & 1 & 0 \\ \dfrac{a_{2,2}}{a_{2,1}} & 0 & 1 \end{bmatrix} = \begin{bmatrix} 1 & 0 & 0 \\ 0 & 1 & 0 \\ \dfrac{139}{-6} & 0 & 1 \end{bmatrix} = \begin{bmatrix} 1 & 0 & 0 \\ 0 & 1 & 0 \\ -23.167 & 0 & 1 \end{bmatrix}
$$

(5.145)

from hence, using (5.72) we obtain

$$
\mathbf{G} = \mathbf{PAP}^{-1} = \begin{bmatrix} 1 & 0 & 0 \\ 0 & 1 & 0 \\ -23.167 & 0 & 1 \end{bmatrix} \begin{bmatrix} 0 & 1 & 0 \\ 0 & 0 & 1 \\ -169 & 38 & 3 \end{bmatrix}
$$

$$
\cdot \begin{bmatrix} 1 & 0 & 0 \\ 0 & 1 & 0 \\ -23.167 & 0 & 1 \end{bmatrix}^{-1} = \begin{bmatrix} 0 & 1 & 0 \\ 23.167 & 0 & 1 \\ 0 & 14.833 & 6 \end{bmatrix}.
$$

(5.146)

It can be easily seen that the result obtained is identical to that obtained in example 5.8.

Example 5.10

Simplify model $K_7(s)$

$$
K_7(s) = \frac{1}{(1+0.1s)^4(s^2+10s+29)(s+1)}
$$

(5.147)

to models $K_6(s)$, $K_5(s)$, $K_4(s)$ of orders $k = 6$, 5 and 4 respectively, by means of comparison of the characteristic equation coefficients.

Solution

The denominator $M_7(s)$ of the model $K_7(s)$ can be written in the form

$$
M_7(s) = 10^{-4}s^7 + 5.1\cdot 10^{-3}s^6 + 0.1079s^5 + 1.2189s^4
$$

$$
+ 7.856s^3 + 28.35s^2 + 50.60s + 29.
$$

(5.148)

We check the possibility of simplification $K_7(s)$ by the order $k = 6$ inserting $s = \dfrac{6S}{50.60}$ into $M_7(s)$. We obtain

$$
M_7(S) = 3.2961\cdot 10^{-11}S^7 + 1.4176\cdot 10^{-8}S^6 + 2.5294\cdot 10^{-6}S^5
$$

$$
+ 2.4097\cdot 10^{-4}S^4 + 1.3097\cdot 10^{-2}S^3 + 0.3984S^2 + 6.000S + 29.
$$

(5.149)

As the coefficient at S^7 is significantly lower than 1, we can neglect the expression $10^{-4}s^7$ in the denominator $M_7(s)$, thus obtaining a model being one order lower, with the denominator $M_6(s)$

$$M_6(s) = 5.1 \cdot 10^{-3} s^6 + 0.1079 s^5 + 1.2189 s^4$$
$$+ 7.856 s^3 + 28.35 s^2 + 50.60 s + 29 . \qquad (5.150)$$

Repeating the procedure for $k = 5$ we insert $s = \dfrac{5S}{50.60}$ into $M_6(s)$ and obtain

$$M_6(S) = 4.7477 \cdot 10^{-9} S^6 + 1.0165 \cdot 10^{-6} S^5 \qquad (5.151)$$
$$+ 1.1621 \cdot 10^{-4} S^4 + 0.007579 S^3 + 0.2767 S^2 + 5S + 29 .$$

As the coefficient at S^6 of this model is also lower than 1 we neglect expression $5.1 \cdot 10^{-3} s^6$ in the denominator $M_6(s)$ and our model is simplified to the order of 5 with the denominator $M_5(s)$

$$M_5(s) = 0.1079 s^5 + 1.2189 s^4 + 7.856 s^3 + 28.35 s^2 + 50.60 s + 29 . \qquad (5.152)$$

It can be easily shown that proceeding in an analogous way we can also simplify the denominator $M_5(s)$ obtaining $M_4(s)$ of the fourth order

$$M_4(s) = 1.2189 s^4 + 7.856 s^3 + 28.35 s^2 + 50.60 s + 29 . \qquad (5.153)$$

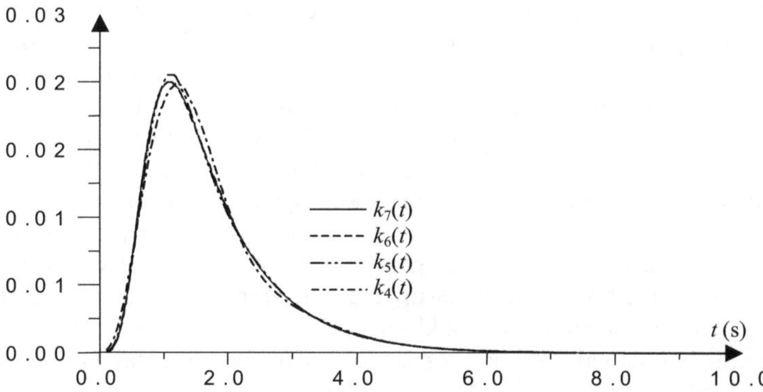

Fig. 5.8. Impulse responses of the primary model $K_7(s)$ and the simplified models $K_6(s)$, $K_5(s)$ and $K_4(s)$ of denominators $M_6(s)$ (5.150), $M_5(s)$ (5.152) and $M_4(s)$ (5.153) respectively.

6. MAXIMUM MAPPING ERRORS

The success of any simulation experiment depends on the accuracy of the model upon which the experiment is performed. This accuracy is verified by the determination of model errors that are defined in different function spaces, by means of chosen error functionals. The determination of errors is possible on the precondition resulting from a previous assumption of a determined input signal used for computations. Traditionally, the dynamic errors are computed for standard input signals, most often in the form of the unit step function, Dirac's impulse or less often of ramp or sinosoidal inputs. There are well established and commonly used methods of computing the various error criteria for these signals, among which the most popular are the integral-square-error and absolute error criterions. As a result, different error values are obtained, since they depend essentially on the input signal for which they are computed. This is a significant limitation of their usefulness, because in practice real systems are not excited by standard signals, but usually by unknown dynamic signals, which are decidedly different from the standard ones. For that reason, the determination of such errors, whose values will be valid for arbitrary dynamic signals, which can appear at the input of the investigated system, is of essential importance. It should be stressed that these can be non-determined signals whose form we are not able to predict a priori. It can be noted that the solution of a problem posed in such a way, which could make the error values independent of the input signal form, is possible for maximum errors. However the procedure of determination of maximum errors requires special input signals to be used, which warrant that the error values determined with them will always be higher or at least equal to the value generated by any other signal. Shapes of signals which maximise the chosen error criteria can be determined based on analytical solution which gives precise results or, in case of non-existence of such solutions, by means of approximated methods using numerical calculations and special computer programs. It is worth stressing however, that in the majority of cases we will have to deal with the latter case, as analytical solutions are most often difficult or impossible to obtain. In the Examples we will present analytical methods which permit the determination of signals that maximise the value of the absolute error and integral-square-error for inputs with a single constraint. Numerical methods for signals with two constraints and integral-square-error criterion will be also demonstrated.

6.1. Input signals with one constraint

6.1.1. Maximum value of the integral-square-error

Suppose the mapping error of the primary model by the simplified one can be expressed by means of the integral-square-error

$$I_2 = \int_0^T [\varepsilon(t)]^2 dt \tag{6.1}$$

where $\varepsilon(t)$ denotes the difference between the responses of the primary model $y_p(t)$ and the simplified model $y_s(t)$

$$\varepsilon(t) = y_p(t) - y_s(t). \tag{6.2}$$

Let us determine the input signal, which maximises the functional (6.1) [40], [43], [45]. For this purpose let us write the responses $y_p(t)$ and $y_s(t)$ by means of convolution integrals

$$y_p(s) = \int_0^t k_p(t-\tau)u(\tau)d\tau \tag{6.3}$$

$$y_s(s) = \int_0^t k_s(t-\tau)u(\tau)d\tau \tag{6.4}$$

in which $k_p(t)$ and $k_s(t)$ represent the impulse responses of the primary and simplified model respectively. Let us denote the difference between $k_p(t)$ and $k_s(t)$ by $k(t)$

$$k(t) = k_p(t) - k_s(t) \tag{6.5}$$

and present the error (6.1) by means of an inner product

$$I_2(u) = (Ku, Ku). \tag{6.6}$$

It can also be given in the equivalent form

$$I_2(u) = (K^* Ku, u) \qquad (6.7)$$

where Ku presents the convolution integral

$$Ku(t) = \varepsilon(t) = \int_0^t k(t-\tau)\, u(\tau)\, d\tau \qquad (6.8)$$

and the K^* is an operator which is conjugate to K. Let us determine the form of the operator K^*. Making use of the properties of inner products and of relationships (6.6) and (6.8), we can write formula (6.1) in the following way

$$I_2(u) = (\varepsilon, Ku) = (u, K^* \varepsilon) \qquad (6.9)$$

that is

$$\int_0^T \varepsilon(t) \int_0^t k(t-\tau) u(\tau)\, d\tau\, dt = \int_0^T u(t)[K^* \varepsilon]\, dt . \qquad (6.10)$$

Extending the integration interval of the internal integral of the left hand side of equation (6.10) up to [0,T] and simultaneously multiplying it by $\mathbf{1}(t-\tau)$, we obtain

$$\int_0^T \varepsilon(t) \int_0^T k(t-\tau)\mathbf{1}(t-\tau) u(\tau)\, d\tau\, dt = \int_0^T u(t)[K^* \varepsilon]\, dt \qquad (6.11)$$

which, after changing the integration order and replacing t by τ, gives

$$\int_0^T u(t) \int_0^T k(\tau-t)\mathbf{1}(\tau-t)\varepsilon(\tau)\, d\tau\, dt = \int_0^T u(t)[K^* \varepsilon]\, dt . \qquad (6.12)$$

Taking into account that the integrand in (6.12) becomes zero for $\tau < t$, we can write this equation in the form

$$\int_0^T u(t) \int_t^T k(\tau-t)\varepsilon(\tau)\, d\tau\, dt = \int_0^T u(t)[K^* \varepsilon]\, dt \qquad (6.13)$$

from whence it results that

$$K^*\varepsilon = \int_t^T k(\tau-t)\varepsilon(\tau)d\tau \ . \tag{6.14}$$

Hence

$$K^*Ku = \int_t^T k(\tau-t)\left(\int_0^\tau k(\tau-v)u(v)dv\right)d\tau . \tag{6.15}$$

Let the signal $u(t)$ maximising $I_2(u)$ be denoted by $u_0(t)$, producing

$$I_2(u_0) = sup\{\ I_2(u): \ u\in U\ \} \tag{6.16}$$

where, the set U is assumed to contain signals being measurable in the meaning of Lebesgue. Let the signals $u\in U$ be limited in magnitude

$$|u(t)|\le a \qquad 0<a\le 1 . \tag{6.17}$$

From the condition of optimality (6.16) it results that

$$(\frac{\partial I_2(u)}{\partial u}\Big|_{u_0},u-u_0)\le 0 . \tag{6.18}$$

Having computed the derivative $\dfrac{\partial I_2(u)}{\partial u}\Big|_{u_0}$, considering (6.7) and performing simple transformations (6.18), yields

$$(K^*Ku_0,u)\le (K^*Ku_0,u_0) . \tag{6.19}$$

It can be easily seen that the left hand side of formula (6.19) reaches maximum, making both sides equal if a signal with a maximum permissible magnitude a

$$|u_0(t)| = a = 1 \tag{6.20}$$

has the form

$$u(t) = u_0(t) = sign[K^* K u_0(t)].$$

(6.21)

After considering (6.15) we finally obtain

$$u_0(t) = sign\left[\int_t^T k(\tau - t)\left(\int_0^\tau k(\tau - v)u_0(v)dv\right)d\tau\right].$$

(6.22)

The maximum value of the error $max\ I_2(u)$ generated by the signal $u_0(t)$ is

$$max\ I_2(u) = I_2(u_0) = \int_0^T \left|K^* K u_0(t)\right| dt$$

$$= \int_0^T \left|\int_t^T k(\tau - t)\left(\int_0^\tau k(\tau - v)u_0(v)dv\right)d\tau\right| dt.$$

(6.23)

6.1.2. Algorithm for determining the maximum value of the integral-square-error

From formula (6.22) it results that $u_0(t)$ is a signal of the "bang–bang" type with maximum magnitude assuming the value of +1 or −1 by virtue of formula (6.20) and with the switching instants t_1, t_2, ... t_n corresponding to the consecutive $i = 1, 2, ...n$ zeros of the function occurring under the *sign* mark in formula (6.22). In order to determine these instants, let us assume that the first switching of the signal $u_0(t)$ occurs from +1 to −1. This means that in the first time interval for $0 < t < t_1$ signal $u_0(t) = +1$. Let us also assume that we will search for n switchings over interval [0,T]. On the basis of formula (6.22) we can write n equations with $t_1, t_2, ... t_n$ as variables for those assumptions. It can be easily seen that the equations are described by the following relationship [45]:

$$\sum_{l=i}^n \int_{t_l}^{t_{l+1}} k(\tau - t_i)\left(\sum_{m=0}^l (-1^m)\int_{t_m}^{t_{m+1}} k(\tau - v)dv\right)d\tau = 0 \qquad i = 1, 2, ...n$$

(6.24)

where $t_0 = 0$, $t_{n+1} = T$, $t_{m+1} = \tau$ for $m = l$ n - number of switchings. Solution of equation system (6.24) with respect to $t_1, t_2, ... t_n$ gives the sought switchings instants of the signal $u_0(t)$. Between those instants, depending on the interval $t < t_1$, $t_1 \le t < t_2$, ... $t \ge t_n$, function $K^* K u_0(t, t_1, \ldots t_n)$ (6.21) is determined by the system of $n+1$ following relationships

$$K^* K u_0(t, t_1, t_2, \ldots t_n) = \sum_{l=i-1}^{n} \int_{t_l}^{t_{l+1}} k(\tau - t) \left(\sum_{m=0}^{l} (-1)^m \int_{t_m}^{t_{m+1}} k(\tau - v) dv \right) d\tau \qquad (6.25)$$

$$i = 1, 2, \ldots n+1$$

where $t_0 = 0$, $t_{n+1} = T$, $t_l = t$ for $l = i-1$, $t_{m+1} = \tau$ for $m = l$, n - number of switchings. The value $I_2(u_0)$ is determined by the sum of modules, which is determined by formula (6.25) over all $n+1$ intervals

$$I_2(u_0) = \sum_{i=1}^{n+1} \left| \sum_{l=i-1}^{n} \int_{t_l}^{t_{l+1}} k(\tau - t) \left(\sum_{m=0}^{l} (-1)^m \int_{t_m}^{t_{m+1}} k(\tau - v) dv \right) d\tau \right|. \qquad (6.26)$$

The procedure of searching for the optimum number of n switchings commences with the assumption $i = 1$, the solution of equation (6.24) with respect to t_1, and after checking the value $I_2(u_0)$ (6.26) corresponding to the obtained solution. Next the procedure is repeated for $i = 2, 3, \ldots$. In this way, the upper value of the index n is not given in advance, but it is being consecutively increased until the $I_2(u_0)$ resulting from formula (6.26) reaches a maximum. Such a situation occurs when the value $I_2(u_0)$ reached for $n+1$ switchings is not higher than the value of this functional corresponding to n switchings, and a further increase in the number of switchings cannot increase it any further. In consequence the process of searching for the optimum number of switchings will conclude at this value of n.

As an example, we will present equations below which result from formulae (6.24) and (6.25) for switching instants at t_1 and t_2 as well as in t_1, t_2 and t_3 correspondingly. Searching for two switchings at t_1 and t_2 using formula (6.24) we obtain two equations resulting from (6.24):

$$\int_{t_1}^{t_2} k(\tau - t_1) \left[\int_0^{t_1} k(\tau - v) dv - \int_{t_1}^{\tau} k(\tau - v) dv \right] d\tau$$

$$+ \int_{t_2}^{T} k(\tau - t_1) \left[\int_0^{t_1} k(\tau - v) dv - \int_{t_1}^{t_2} k(\tau - v) dv + \int_{t_2}^{\tau} k(\tau - v) dv \right] d\tau = 0 \qquad (6.27)$$

and

$$\int_{t_2}^{T} k(\tau - t_2)\left[\int_{0}^{t_1} k(\tau - v)dv - \int_{t_1}^{t_2} k(\tau - v)dv + \int_{t_2}^{\tau} k(\tau - v)dv\right]d\tau = 0 \tag{6.28}$$

and we have three equations resulting from (6.25):

for $t < t_1$

$$K^* K u_0(t,t_1,t_2) = \int_{t}^{t_1} k(\tau - t)\left[\int_{0}^{\tau} k(\tau - v)dv\right]d\tau$$

$$+ \int_{t_1}^{t_2} k(\tau - t)\left[\int_{0}^{t_1} k(\tau - v)dv - \int_{t_1}^{\tau} k(\tau - v)dv\right]d\tau \tag{6.29}$$

$$+ \int_{t_2}^{T} k(\tau - t)\left[\int_{0}^{t_1} k(\tau - v)dv - \int_{t_1}^{t_2} k(\tau - v)dv + \int_{t_2}^{\tau} k(\tau - v)dv\right]d\tau$$

for $t_1 \le t_2$

$$K^* K u_0(t,t_1,t_2) = \int_{t}^{t_2} k(\tau - t)\left[\int_{0}^{t_1} k(\tau - v)dv - \int_{t_1}^{\tau} k(\tau - v)dv\right]d\tau$$

$$+ \int_{t_2}^{T} k(\tau - t)\left[\int_{0}^{t_1} k(\tau - v)dv - \int_{t_1}^{t_2} k(\tau - v)dv + \int_{t_2}^{\tau} k(\tau - v)dv\right]d\tau \tag{6.30}$$

and for $t \ge t_2$

$$K^* K u_0(t,t_1,t_2) = \int_{t}^{T} k(\tau - t)\left[\int_{0}^{t_1} k(\tau - v)dv\right.$$

$$\left. - \int_{t_1}^{t_2} k(\tau - v)dv + \int_{t_2}^{\tau} k(\tau - v)dv\right]d\tau. \tag{6.31}$$

For three switchings at t_1, t_2 and t_3 we have three equations resulting from (6.24):

$$\int_{t_1}^{t_2} k(\tau - t_1) \left[\int_0^{t_1} k(\tau - v)dv - \int_{t_1}^{\tau} k(\tau - v)dv \right] d\tau$$

$$+ \int_{t_2}^{t_3} k(\tau - t_1) \left[\int_0^{t_1} k(\tau - v)dv - \int_{t_1}^{t_2} k(\tau - v)dv + \int_{t_2}^{\tau} k(\tau - v)dv \right] d\tau \qquad (6.32)$$

$$+ \int_{t_3}^{T} k(\tau - t_1) \left[\int_0^{t_1} k(\tau - v)dv - \int_{t_1}^{t_2} k(\tau - v)dv + \int_{t_2}^{t_3} k(\tau - v)dv - \int_{t_3}^{\tau} k(\tau - v)dv \right] d\tau = 0$$

$$\int_{t_2}^{t_3} k(\tau - t_2) \left[\int_0^{t_1} k(\tau - v)dv - \int_{t_1}^{t_2} k(\tau - v)dv + \int_{t_2}^{\tau} k(\tau - v)dv \right] d\tau$$

$$+ \int_{t_3}^{T} k(\tau - t_2) \left[\int_0^{t_1} k(\tau - v)dv - \int_{t_1}^{t_2} k(\tau - v)dv + \int_{t_2}^{t_3} k(\tau - v)dv - \int_{t_3}^{\tau} k(\tau - v)dv \right] d\tau = 0 \qquad (6.33)$$

$$\int_{t_3}^{T} k(\tau - t_3) \left[\int_0^{t_1} k(\tau - v)dv - \int_{t_1}^{t_2} k(\tau - v)dv + \int_{t_2}^{t_3} k(\tau - v)dv - \int_{t_3}^{\tau} k(\tau - v)dv \right] d\tau = 0 \qquad (6.34)$$

and four equations resulting from (6.25):

for $t < t_1$

$$K^* K u_0(t, t_1, t_2, t_3) = \int_t^{t_1} k(\tau - t) \left[\int_0^{\tau} k(\tau - v)dv \right] d\tau$$

$$+ \int_{t_1}^{t_2} k(\tau - t) \left[\int_0^{t_1} k(\tau - v)dv - \int_{t_1}^{\tau} k(\tau - v)dv \right] d\tau$$

$$+ \int_{t_2}^{t_3} k(\tau - t) \left[\int_0^{t_1} k(\tau - v)dv - \int_{t_1}^{t_2} k(\tau - v)dv + \int_{t_2}^{\tau} k(\tau - v)dv \right] d\tau \qquad (6.35)$$

$$+ \int_{t_3}^{T} k(\tau - t) \left[\int_0^{t_1} k(\tau - v)dv - \int_{t_1}^{t_2} k(\tau - v)dv + \int_{t_2}^{t_3} k(\tau - v)dv - \int_{t_3}^{\tau} k(\tau - v)dv \right] d\tau$$

for $t_1 \le t < t_2$

$$K^* K u_0(t,t_1,t_2,t_3) = \int\limits_t^{t_2} k(\tau-t)\left[\int\limits_0^{t_1} k(\tau-v)dv - \int\limits_{t_1}^{\tau} k(\tau-v)dv\right]d\tau$$

$$+ \int\limits_{t_2}^{t_3} k(\tau-t)\left[\int\limits_0^{t_1} k(\tau-v)dv - \int\limits_{t_1}^{t_2} k(\tau-v)dv + \int\limits_{t_2}^{\tau} k(\tau-v)dv\right]d\tau \qquad (6.36)$$

$$+ \int\limits_{t_3}^{T} k(\tau-t)\left[\int\limits_0^{t_1} k(\tau-v)dv - \int\limits_{t_1}^{t_2} k(\tau-v)dv + \int\limits_{t_2}^{t_3} k(\tau-v)dv - \int\limits_{t_3}^{\tau} k(\tau-v)dv\right]d\tau$$

for $t_2 \le t < t_3$

$$K^* K u_0(t,t_1,t_2,t_3) = \int\limits_t^{t_3} k(\tau-t)\left[\int\limits_0^{t_1} k(\tau-v)dv - \int\limits_{t_1}^{t_2} k(\tau-v)dv + \int\limits_{t_2}^{\tau} k(\tau-v)dv\right]d\tau$$

$$+ \int\limits_{t_3}^{T} k(\tau-t)\left[\int\limits_0^{t_1} k(\tau-v)dv - \int\limits_{t_1}^{t_2} k(\tau-v)dv + \int\limits_{t_2}^{t_3} k(\tau-v)dv - \int\limits_{t_3}^{\tau} k(\tau-v)dv\right]d\tau \qquad (6.37)$$

for $t > t_3$

$$K^* K u_0(t,t_1,t_2,t_3) = \int\limits_t^{T} k(\tau-t)\left[\int\limits_0^{t_1} k(\tau-v)dv\right.$$

$$\left. - \int\limits_{t_1}^{t_2} k(\tau-v)dv + \int\limits_{t_2}^{t_3} k(\tau-v)dv - \int\limits_{t_3}^{\tau} k(\tau-v)dv\right]d\tau. \qquad (6.38)$$

For a higher number of switchings $t_1, t_2 \ldots t_n$, we can set up a relevant system of any n equations in a similar way as for $n = 3$. In the case where equation (6.24) has an equivocal solution, then all possible combinations of the obtained solutions should be checked, verifying them for the sums of absolute values using formula (6.26). This occurs when in the fixed integration interval [0,T], the equation (6.24) becomes equal to zero more times than the assumed number n. The maximum value of the sum obtained for the assumed number of n switchings determines the solution being sought.

6.1.3. Maximum value of absolute error.

Let us consider a similar problem for the criterion of the maximum value of error D presented by formula

$$D = |\varepsilon(t)| \tag{6.39}$$

where the error $\varepsilon(t)$ is given by (6.2).

Let us determine the input signal $u(t) = u_0(t)$, which maximises error D over the interval $[0,T]$. Thus we have

$$D(u_0) = \max_{t\in[0,T]} \max_{u\in U} D(u). \tag{6.40}$$

Representing the error $\varepsilon(t)$ in the form of a convolution integral, we can write the right hand side of equation (6.40) in the form

$$\max_{t\in[0,T]} \max_{u\in U} |\varepsilon(t)| = \max_{t\in[0,T]} \max_{u\in U} \left| \int_0^t k(t-\tau)u(\tau)\,d\tau \right| \tag{6.41}$$

where the impulse response $k(t)$ is determined by (6.5). From formula (6.41) it results that error $\varepsilon(t)$ reaches maximum value at the instant $t = T$ if the input signal with maximum magnitude (6.20) satisfies the condition

$$u(t) = u_0(t) = \operatorname{sign} k(T-t). \tag{6.42}$$

The maximum error value which corresponds to the signal $u_0(t)$ is

$$D(u_0) = \int_0^T |k(t)|\,dt \tag{6.43}$$

which is not difficult to compute.

6.2. Input signals with two constraints

Mapping errors of models being determined by means of "bang–bang" signals are at maximum due to their determination procedure. For this reason, errors determined using these signals reach relatively high values even when the simplified model represents the model of the system quite well. This is caused by the particular dynamics of the "bang–bang" signals, which have infinitely high derivative values in the instants of switching. Outside of these instants, they have constant values. The dynamics of these signals is not matched to the dynamics of physically existing systems, since they can only transmit signals with limited value of rate of change. In order to match the dynamic behaviour of the input signals we will impose a constraint resulting from the dynamic properties of the system being modelled in addition to magnitude constraint (6.17). Proper matching is obtained by restricting the maximum rate of change of the input signal $\max|u'(t)|$ to a value less or at most equal to the maximum rate of the step response of the system. Denoting this rate of change by ϑ we can write

$$\vartheta = \max_{t\in[0,T]} \left| u'(t) \right| \le \max_{t\in[0,\infty]} \left| h'(t) \right| = \max_{t\in[0,\infty]} \left| k(t) \right| \tag{6.44}$$

where $h(t)$ and $k(t)$ denote the step and impulse responses of the system. The constraint ϑ can be also determined using the frequency response of the system. Let us denote the maximum angular frequency corresponding to the cut-off frequency of this response by ω_m. This means from the assumption, that the series of harmonics

$$u(t) = A\sin\omega t \qquad \omega = 1, 2, ...\omega_m \tag{6.45}$$

up to the frequency ω_m inclusive, should be transmitted by this system without deformations. The maximum velocity of the signal corresponding to the highest harmonic of this series is thus

$$\vartheta = \max\left| u'(t) \right| = A\omega_m \tag{6.46}$$

where the limit ω_m is arbitrary and can, but does not need to, correspond to a 3dB reduction in signal magnitude.

6.2.1. Domain of signals maximising the integral-square-error

Suppose that, following the analysis of the investigated system dynamic properties, we have established the values of constraints a (6.17) and ϑ (6.44). Our problem is now reduced to the determination of the $u_0(t)$ signal, which maximises the integral-square-error criterion (6.1) within these constraints. However, an analytical solution such as that given by formula (6.22) for a signal restricted only in magnitude is not known and it is not certain whether this is obtainable. Therefore we will determine the $u_0(t)$ signal by means of numerical computations. Various computer programs can be used for this purpose, however it is essential that the space in which the possible solutions are searched for and in which the $u_0(t)$ signal is, be maximally limited. In order to limit this space we will prove that the signal meeting constraints (6.17) and (6.44) must, over the interval [0,T], always reach the constraint of the magnitude a or the constraint of the rate of change ϑ [46].

Theorem

Suppose U is a set of functions C^1 in sections over the interval $[0,T]$. Let $I_2(u)$ be the positive square functional on U (6.1)

$$I_2(u) = \int_0^T [\varepsilon(t)]^2 dt \geq 0 \tag{6.47}$$

which means that the bilinear symmetrical transformation $\exists \tilde{I}_2$ exists

$$I_2(u) = \tilde{I}_2(Ku, Ku) \qquad u \in U . \tag{6.48}$$

Let the condition be fulfilled

$$\forall\, 0 < b < c < T \qquad \exists\, h \in U : supp\, h \subset [b,c] \tag{6.49}$$

such that

$$I_2(h) \neq 0 \tag{6.50}$$

which automatically means that $I_2(h) > 0$.

Let us define the following set A

$$A = \left\{ u(t) \in U : \; |u(t)| \le a \quad |u'_+(t)| \le \vartheta \quad |u'_-(t)| \le \vartheta \quad t \in [0,T] \right\}. \tag{6.51}$$

Let $u_0(t) \in A$ fulfil the condition

$$I_2(u_0) = \sup\{I_2(u): \; u \in A\} \tag{6.52}$$

then

$$\forall t \in [0,T] \quad |u_0(t)| = a \quad \text{or} \quad |u'_+(t)| = \vartheta \quad \text{or} \quad |u'_-(t)| = \vartheta. \tag{6.53}$$

Proof

Suppose that (6.53) is not true. Then

$$\exists \varepsilon > 0 \quad \exists\, 0 < b < c < T \tag{6.54}$$

such that

$$|u(t)| \le a - \varepsilon, \quad |u'_{0+}(t)| \le \vartheta - \varepsilon, \quad |u'_{0-}(t)| \le \vartheta - \varepsilon. \tag{6.55}$$

Let us choose h according to (6.49)

$$supp\, h \subset [b,c] \qquad I_2(h) > 0 \tag{6.56}$$

then for the small $d \in \Re$, say $d \in (-\delta,\delta)$ is

$$u_0 + dh \in A \qquad \forall d \in (-\delta,\delta) \tag{6.57}$$

and from the optimum condition in $u_0(t)$ it results that

$$I_2(u_0) \geq I_2(u_0 + dh) = I_2(u_0) + 2d\widetilde{I}_2(u_0,h) + d^2 I_2(h), \quad d \in (-\delta,\delta) \qquad (6.58)$$

hence

$$0 \geq 2d\widetilde{I}_2(u_0,h) + d^2 I_2(h) \qquad d \in (-\delta,\delta). \qquad (6.59)$$

However, it can be easy seen that solution (6.59) represents a parabola crossing zero and directed upwards, so that the last inequality will never be fulfilled for $I_2(h) > 0$, $d \in (-\delta,\delta)$.

Corollary

As relationship (6.59) represents a contradiction, so the functional $I_2(u_0)$ can fulfil condition (6.52) iff the input signal $u_0(t)$ reaches one of the constraints given in (6.53).

Carrying out the proof in an identical way to that above, it can be shown that if only one of the constraints is applied to the signal, either of magnitude a or of the rate of change ϑ, then the functional $I_2(u_0)$ reaches maximum iff the signal reaches this constraint over the interval [0, T].

It is worth stressing that the proof presented above is especially useful in situations where we can waive the requirement of an optimum solution in favour of the approximate one, that is when we decide to use heuristic techniques of searching within a space of solutions for an estimated value. One possibility is to use a program based on genetic algorithms. This allows the space to be restricted to a double or quadruple element set, which significantly increases the likelihood of obtaining a correct solution, and considerably reduces the computation time. In cases using the constraint (6.17), the $u_0(t)$ input has the form of a "bang-bang" signal with the magnitude $a = 1$ and the program determines its switching instants only. In cases where there are two simultaneous constraints (6.53), signal $u_0(t)$ can be only in the form of triangles with the slope inclination ϑ_+ or ϑ_- or of trapezoids with the slopes ϑ_+ and ϑ_- and a magnitude of $a = 1$

6.3. Examples

Example 6.1

Determine the characteristic of the mapping error $I_2(u_0) = f(T)$ of a sixth order model from example 5.4 by using the second-order model calculated using formulae (4.50) and (4.42). Assume a signal $u_0(t)$ of the "bag-bang" type with a constraint of magnitude (6.20) and the range of the determined characteristic corresponding to the steady-state of the impulse responses of the models.

Solution

Applying formulae (4.50), (4.42) and the calculation procedure shown in example 5.4 for sixth order model

$$K_6(s) = \frac{6s^4 + 50s^3 + 196s^2 + 418s + 434}{s^6 + 12s^5 + 71s^4 + 256s^3 + 575s^2 + 804s + 585} \tag{6.60}$$

we obtain the following second-order model

$$K_2(s) = \frac{6}{s^2 + 3.66s + 7.78} . \tag{6.61}$$

The impulse responses of the $K_6(s)$ and $K_2(s)$ models in the form of a power series are as follows

$$k_6(t) = \sum_{k=0}^{\infty} c_{6,k} t^k = 6.00t - 11.00t^2 + 5.66t^3 + 1.50t^4 - 1.91t^5 - 0.96t^6 \tag{6.62}$$
$$+ 1.95t^7 - 1.15t^8 + 0.32t^9 - \ldots$$

$$k_2(t) = \sum_{k=0}^{\infty} c_{6,k} t^k = 6.00t - 11.00t^2 + 5.66t^3 + 1.93t^4 - 3.63t^5 + 1.70t^6 \tag{6.63}$$
$$- 0.217t^7 - 0.137t^8 + 0.079t^9 - \ldots .$$

It is easy to see that the first four elements in the corresponding series have the same values. In order to determine characteristic $I_2(u_0)$ let us assume that the value of the upper limit of integration in integral (6.1) is changed at $\Delta T = 1$s. from

$T = 1$s. up to $T = 8$s., which corresponds to the steady state of the difference of impulse responses $k_6(t)$ and $k_2(t)$ - Fig.6.3. It is possible to determine the switching instants of the signal $u_0(t)$ and the values of mapping error $I_2(u_0) = f(T)$ resulting from relationship (6.26) by solving the equation systems resulting from (6.24), for each of the intervals $[0,T]$. The following results have been obtained for the current example:

$$I_2(u_{0,1}) = 0.20 \cdot 10^{-6} [\text{V}^2 s]$$

$$for:$$

$$u_{0,1}(t) \Rightarrow t_0 = 0s., \quad t_1 = 0.26s., \quad t_2 = 0.82s., \quad t_3 = T = 1.0s.$$

(6.64)

$$I_2(u_{0,2}) = 44.02 \cdot 10^{-6} [\text{V}^2 s]$$

$$for:$$

$$u_{0,2}(t) \Rightarrow t_0 = 0s., \, t_1 = 0.4s., \, t_2 = 1.2s., \, t_3 = 1.9s., \, t_4 = T = 2.0s.$$

(6.65)

$$I_2(u_{0,3}) = 13.01 \cdot 10^{-4} [\text{V}^2 s]$$

$$for:$$

$$u_{0,3}(t) \Rightarrow t_0 = 0s., \, t_1 = 1.2s., \, t_2 = 2.2s., \, t_3 = 2.9s., \, t_4 = T = 3.0s.$$

(6.66)

$$I_2(u_{0,4}) = 30.12 \cdot 10^{-4} [\text{V}^2 s]$$

$$for$$

$$u_{0,4}(t) \Rightarrow t_0 = 0s., \, t_1 = 2.0s., \, t_2 = 3.1s., \, t_3 = 3.5s., \, t_4 = T = 4.0s.$$

(6.67)

$$I_2(u_{0,5}) = 60.83 \cdot 10^{-4} [\text{V}^2 s]$$

$$for:$$

$$u_{0,5}(t) \Rightarrow t_0 = 0s., \quad t_1 = 1.4s., \quad t_2 = 3.1s., \quad t_3 = 4.1s., \quad t_4 = 4.8s.,$$

$$t_5 = T = 5.0s.$$

(6.68)

$$I_2(u_{0,6}) = 86.54 \cdot 10^{-4} [\text{V}^2 s]$$

$$for:$$

$$u_{0,6}(t) \Rightarrow t_0 = 0s., \quad t_1 = 1.1s., \quad t_2 = 2.6s., \quad t_3 = 4.2s., \quad t_4 = 5.2s.,$$

$$t_5 = T = 6.0s.$$

(6.69)

$$I_2(u_{0,7}) = 11.73 \cdot 10^{-3}[\mathrm{V}^2 s]$$

for :

$$u_{0,7}(t) \Rightarrow t_0 = 0s., \quad t_1 = 1.6s., \quad t_2 = 3.3s., \quad t_3 = 5.1s., \quad t_4 = 6.1s.,$$

$$t_5 = T = 7.0s.$$

$$(6.70)$$

$$I_2(u_{0,8}) = 14.71 \cdot 10^{-3}[\mathrm{V}^2 s]$$

for :

$$u_{0,8}(t) \Rightarrow t_0 = 0s., \quad t_1 = 1.2s., \quad t_2 = 2.7s., \quad t_3 = 4.3s., \quad t_4 = 6.2s.,$$

$$t_5 = 7.2s., \quad t_6 = 7.8s., \quad t_7 = T = 8.0s.$$

$$(6.71)$$

where the first switching at t_1 comes from +1 to –1. Fig. 6.1. shows the determined characteristic, Fig.6.2 presents the diagrams of the impulse responses of the models and Fig.6.3 their difference $k_{62}(t)$.

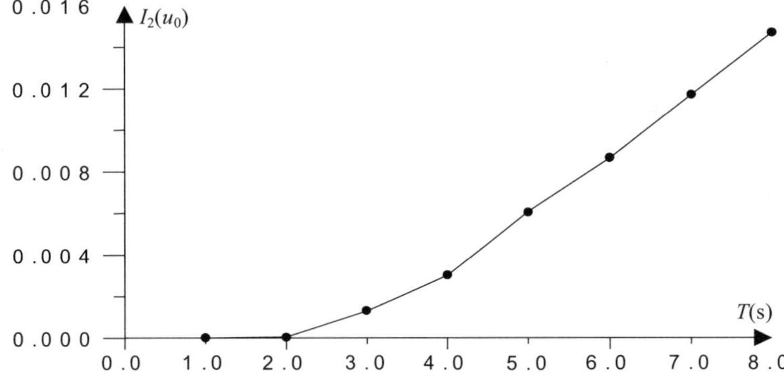

Fig. 6. 1. Characteristic of the error $I_2(u_0) = f(T)$.

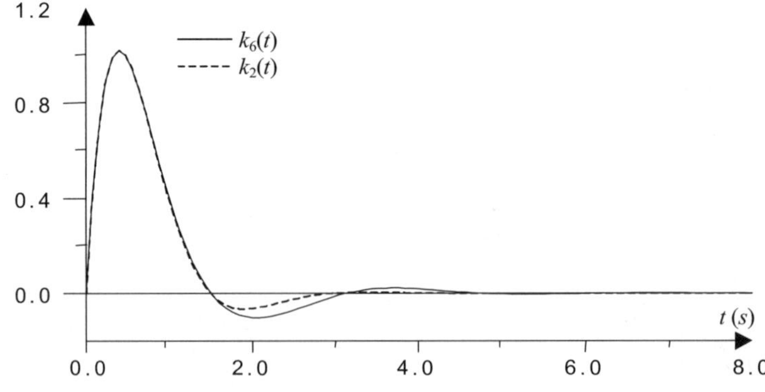

Fig. 6. 2. Impulse responses $k_6(t)$ and $k_2(t)$.

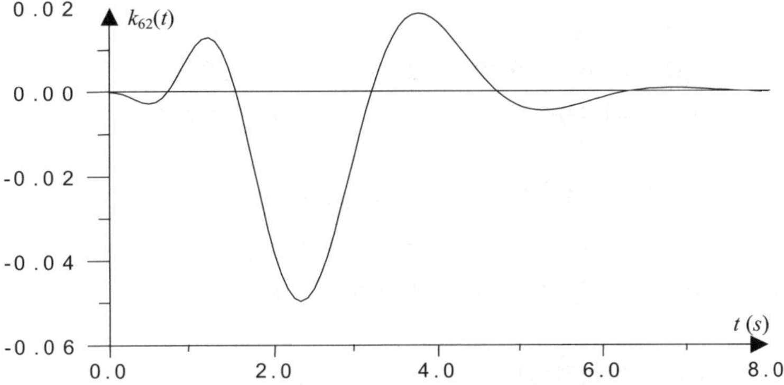

Fig. 6. 3. Difference $k_{62}(t)$ between responses $k_6(t)$ and $k_2(t)$.

Example 6.2

From the previous example determine the maximum value of the mapping error $I_2(u_0)$ for $T = 8$s. for models $K_6(s)$ and $K_2(s)$. Assume that the signal $u_0(t)$ has a constrained magnitude (6.20) and constrained rate of change (6.44) resulting from the impulse response $k_6(t)$ of the model $K_6(s)$.

Solution

We determine the constraints of the signal $u_0(t)$. The constraint of magnitude a is

$$a = \left| u_0(t) \right| = 1 .$$
$$t \in [0,T]$$

(6.72)

The rate of change constraint ϑ is equal to the maximum value of the impulse response $k_6(t)$. It can be easily confirmed that this constraint equals

$$\vartheta = \max_{t \in [0,\infty]} \left| k_6(t) \right| = 1.02$$

(6.73)

and is obtained for $t = 0.405$s. – Fig. 6.2. We determine the signal $u_0(t)$ with applied constraints $a = 1$ and $\vartheta = 1.02$ using the method of genetic algorithms. The program searches the signal $u_0(t)$ which maximises the functional (6.1). Expression $\varepsilon(t)$ is calculated by means of the convolution integral (6.8) where $k(t)$ is the difference of the impulse responses of models $K_6(s)$ and $K_2(s)$.

For five computing cycles with the limit of the maximum of population set at 200, the following signals $u_0(t)$ have been obtained:

$$I_2(u_{0,1}) = 58.41 \cdot 10^{-4} \, [\mathrm{V}^2 s] \quad \textit{No. of population} : 64^{th}$$

$$u_{0,1}(t) \Rightarrow \vartheta_+[0.0,\ 0.96s.],\ \vartheta_-[0.96,\ 2.90s.],\ -1[2.90,\ 2.98s.], \tag{6.74}$$

$$\vartheta_+[2.98, 4.94s.],\ +1[4.94,\ 5.33s.],\ \vartheta_-[5.33, 7.29s.], -1[7.29,\ 8.00s.].$$

$$I_2(u_{0,2}) = 57.87 \cdot 10^{-4} \, [\mathrm{V}^2 s] \quad \textit{No. of population} : 65^{th}$$

$$u_{0,2}(t) \Rightarrow \vartheta_+[0.0,\ 0.98s.],\ +1[0.98,\ 1.55s.],\ \vartheta_-[1.55,\ 3.51s.], \tag{6.75}$$

$$-1[3.51,\ 4.89s.],\ \vartheta_+[4.89, 6.58s.],\ +1[6.58,\ 8.00s.].$$

$$I_2(u_{0,3}) = 58.43 \cdot 10^{-4} \, [\mathrm{V}^2 s] \quad \textit{No. of population} : 10^{th}$$

$$u_{0,3}(t) \Rightarrow \vartheta_-[0.0,\ 0.98s.],\ -1[0.98,\ 1.13s.],\ \vartheta_+[1.13,\ 3.09s.], \tag{6.76}$$

$$\vartheta_-[3.09,\ 5.05s.],\ -1[5.05,\ 5.26s.],\ \vartheta_+[5.26, 7.22s.],\ +1[7.22,\ 8.00s.].$$

$$I_2(u_{0,4}) = 57.38 \cdot 10^{-4} \, [\mathrm{V}^2 s] \quad \textit{No. of population} : 57^{th}$$

$$u_{0,4}(t) \Rightarrow \vartheta_-[0.0,\ 0.98s.],\ -1[0.98,\ 1.59s.],\ \vartheta_+[1.59,\ 3.55s.], \tag{6.77}$$

$$+1[3.55,\ 4.90s.],\ \vartheta_-[4.90, 6.86s.],\ -1[6.86, 7.01s.],\ \vartheta_+[7.01, 8.00s.].$$

$$I_2(u_{0,5}) = 58.03 \cdot 10^{-4} \, [\mathrm{V}^2 s] \quad \textit{No. of population} : 187^{th}$$

$$u_{0,5}(t) \Rightarrow \vartheta_+[0.0,\ 0.98s.],\ +1[0.98,\ 1.58s.],\ \vartheta_-[1.58,\ 3.54s.], \tag{6.78}$$

$$-1[3.54,\ 4.92s.],\ \vartheta_+[4.92, 6.86s.],\ \vartheta_-[6.85, 8.00s.].$$

In the above expressions the following notation is used: $\vartheta_+[.]$ signal increasing in the interval [.], $\vartheta_-[.]$ signal decreasing in the interval [.], $\pm 1[.]$ a constant signal of ± 1 in the interval [.], where the tangent of the rise angle is equal to the tangent of the fall angle and its value is 1.02.

The maximum value of $I_2(u_0)$ was obtained in the third test, so we assume that the mapping error is $I_2(u_{0,3}) = 58.43 \cdot 10^{-4} \, [\mathrm{V}^2 s]$.

7. SIGNALS MAXIMISING THE INTEGRAL-SQUARE-ERROR IN THE PROCESS OF MODELS OPTIMISATION

The previous chapter was dedicated to the determination of maximum mapping errors by means of signals with one constraint or two simultaneous constraints imposed upon them. In this chapter we will consider the problem of minimisation of such errors. It is easy to see that for a given order and type of simplified model minimisation of the maximum mapping error is reduced to the solutions known in the domain of parametric optimisation. However, commonly applied methods of such optimisation refer mainly to the minimisation of a chosen objective function to a standard input signal, which in the majority of cases, is the unit step input. A good illustration of this can be seen through the example in [71] which presents six different optimum second order models of a seventh order real system of a supersonic transport aircraft pitch rate, obtained for six different objective functions and for a common unit step input signal. It should be noted however, that the parameters of the optimum models as well as the values of maximum errors depend equally on the chosen objective function and the input signals applied in the optimisation process. For this reason, the parameters of the optimum models obtained for one of the objective functions using the chosen standard signal can still significantly differ from the model parameters for that objective function with a different input signal. In order to make the optimisation results independent of input signal shape or the minimised mapping errors based on them, the optimisation of simplified models will be accomplished by replacing the standard signals with signals which maximise the chosen error functional. Hence on the same grounds for which we previously determined the maximum values of such errors, the optimisation of simplified models will be realised by replacing the standard signals with signals which maximise the chosen error functional. The task then assumes the form of a parametric optimisation of the *minimax* type and is realised in two stages. At the first stage an input signal is determined which maximises the assumed objective function. At the second stage, a model parameter optimisation is carried out on the minimum value of the objective function for the input signal determined at the first stage. An optimisation carried out in such a way can find broad application in the process of determining optimum models of systems, especially for those that operate in dynamic states e.g. measuring systems intended for dynamic measurement, automation systems for control of dynamic quantities etc. as well as for more simplified models of the above systems. The latter case is shown in block diagram in Fig. 7.1 and a good illustration of its application can be the optimisation of models, which we will consider below using low-pass filters as an example.

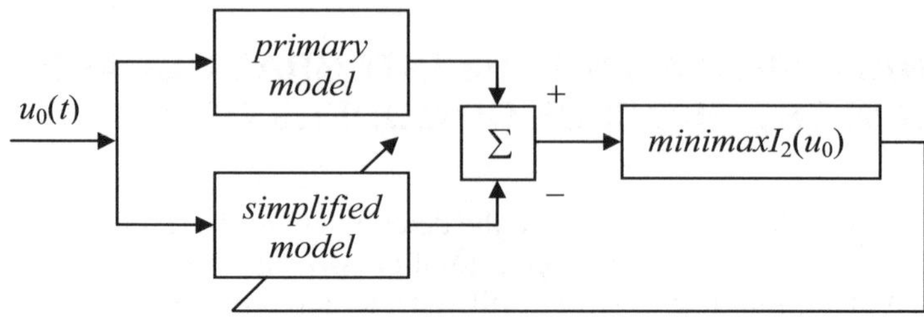

Fig. 7. 1. Block diagram of optimisation process.

7.1. Optimisation of models in the case of the high value of primary mapping error. Optimisation of Butterworth filters

Measuring filters are traditionally optimised in the frequency domain and it seems that there is no good reason to conduct their optimisation in the time domain again. However, this is not the case, as the optimality of measuring filters in the frequency domain does not mean that they are optimal in the time domain and respective correspondence between criteria defined in those two domains exists only in a few cases. Moreover criteria related to frequency responses can assume different forms and their choice is not unambiguously expressed. Most often one of the following criteria is applied:

- maximally flat frequency responses - Butterworth filters,
- uniformly wavy frequency responses - Chebyshev filters,
- frequency independent group time delay - Bessel - Thomson filters.

   Mathematical models of filters which satisfy those requirements have been known for a long time and can be obtained by making use of the coefficients of Chebyshev, Butterworth or Bessel polynomials in the denominators of the respective models, represented by means of transfer functions. In the case of input signals with a known spectral distribution, the signal processing error can be easily determined. The same applies to the mapping error for reducing the filter order or replacing a filter of a higher order with one of a lower order e.g. to reduce the degree of complication of the system where the filter is used or to reduce system production cost, etc. When the spectral distribution of the input signal is not known, there is no possibility of determining those errors and such a situation occurs just in the case of dynamic signals. Below as an example we will carry out an optimisation of some chosen models of Butterworth filters by means of the "bang-bang" signals constrained in magnitude only (6.20) as well as signals with two simultaneous constrains of magnitude (6.20) and of rate of change (6.40). In all cases we will determine the $u_0(t)$ signals by means of a program based on genetic algorithms [23].

7.2. Examples

Example 7.1

Determine the optimum models of filters of the fourth and third order with respect to *minimax* $I_2(u_0)$, referring to the Butterworth filter of the sixth order.

Solution

7.1.1. Input data to the program

Impulse responses of filters

The transfer functions of Butterworth filters of the sixth, fourth and third order are as follows [75]:

$$K_6(s) = \frac{1}{(s^2 + a_1 s + 1)(s^2 + a_2 s + 1)(s^2 + a_3 s + 1)} \tag{7.1}$$

$$a_1 = 1.9319, \quad a_2 = 1.4142, \quad a_3 = 0.5176,$$

$$K_4(s) = \frac{1}{(s^2 + a_1 s + 1)(s^2 + a_2 s + 1)} \tag{7.2}$$

$$a_1 = 1.8478, \quad a_2 = 0.7654,$$

$$K_3(s) = \frac{1}{(s^2 + a_1 s + 1)(a_2 s + 1)} \tag{7.3}$$

$$a_1 = a_2 = 1.$$

Impulse responses of filters given by formulae (7.1)-(7.3) are respectively

$$k_6(t) = 4.5721 \exp(-0.9659t) \sin(0.2587t) + 2.6385 \exp(-0.9659t)$$

$$\cdot \cos(0.2587t) + 5.8436 \cdot 10^{-5} \exp(-0.7071t) \sin(0.7071t)$$

$$- 3.0467 \exp(-0.7071t) \cos(0.7071t) - 3.0467 \exp(-0.7071t) \tag{7.4}$$

$$\cdot \cos(0.701t) - 0.7071 \exp(-0.2588t),$$

$$k_4(t) = 2.2307\exp(-0.9239t)\cdot\sin(0.3826t) + 0.9239$$
$$\cdot\exp(-0.9239t)\cos(0.3826t) - 0.3826\exp(-0.3826t)\sin(0.9239t) \tag{7.5}$$
$$- 0.9239\exp(-0.3826t)\cos(0.9239t),$$

$$k_3(t) = \exp(-t) + 0.5773\exp(-0.5t)\sin(0.8660t) - \exp(-0.5)\cos(0.8660) \tag{7.6}$$

and are represented by diagrams shown in Fig.7.2.

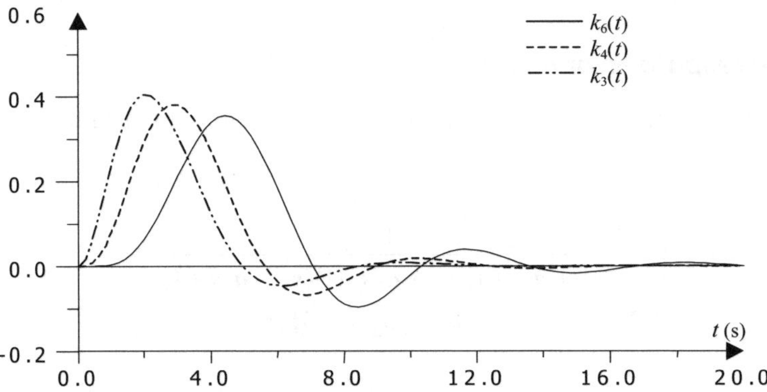

Fig. 7.2. Impulse responses $k_6(t)$, $k_4(t)$ and $k_3(t)$ of Butterworth filters of the sixth, fourth sixth, fourth and third order respectively.

Constraints on input signals

The constraint on magnitude of the $u_0(t)$ signal is (6.20)

$$\left|u_0(t)\right| = 1. \tag{7.7}$$
$$\scriptstyle t\in[0,20]$$

The constraint on rate of change ϑ of the $u_0(t)$ signal, corresponding to the maximum value of the impulse response of the filter $k_6(t)$ (7.4) is (6.44)

$$\vartheta = \max_{t\in[0,\infty]} \left| k_6(t) \right| = 0.355 \tag{7.8}$$

and is reached for $t = 4.44$s.

Upper limit of integration

Two cases of upper limit of integration in (6.1) will be considered. In the first case we will assume that time T corresponds to the steady state value of the impulse response of the sixth order filter. This time is equal to $T = 20$s. and can be read from the diagram shown in Fig. 7.2. In the second case, assuming a high speed dynamics of input signal, we will reduce this time to $T = 10$s.

Range of optimisation

Let us assume that the optimisation will be made between the range a_1 and a_2 for which the second order terms existing in the denominators of the transfer functions of models $K_4(s)$ and $K_3(s)$ will have an oscillation character or will be at the oscillation limit. This means that the dynamic behaviour of these terms after optimisation will be of the same character as before optimisation. These ranges for the $K_4(s)$ filter are: $0 < a_1 < 2$ and $0 < a_2 < 2$ while for the $K_3(s)$ filter they are: $0 < a_1 < 2$ and $a_2 > 0$.

The first stage of the program is commenced using the above data to search for the signal $u(t) = u_0(t)$, which maximises functional (6.1) in which the error $\varepsilon(t)$ is calculated by means of convolution of the signal $u(t)$ and the difference $k_{6,n}(t)$

$$\varepsilon(t) = k_{6n}(t) * u(t) \tag{7.9}$$

$$k_{6n}(t) = k_6(t) - k_n(t) \tag{7.10}$$

where k_n for $n = 4$ and $n = 3$ represent impulse responses of filters given by (7.5) and (7.6) respectively.

In the second stage the program minimises the value of functional $I_2(u_0)$ for the signal $u_0(t)$ to the value of *minimax* $I_2(u_0)$, tuning the parameters of the respective filters.

7.1.2. Results of optimisation

Optimisation of fourth order filter for T = 20s.

Signal $u_0(t)$ with one constraint

The $u_{0,4}(t)$ signal of "bang-bang" type of magnitude $a = 1$, calculated by the program on the base of the difference $k_{64}(t)$ (7.10), giving over [0,20s] the maximum value of the functional $I_2(u_0)$ of 11.692 [V^{2}s] has been obtained for the following switching instants

$$u_{04}(t) \Rightarrow t_0 = 0s., \quad t_1 = 0.483s., \quad t_2 = 4.222s., \quad t_3 = 7.713s., \quad t_4 = 11.199s.,$$
$$t_5 = 15.037s., \quad t_6 = 19.773s., \quad t_7 = T = 20.00s. \tag{7.11}$$

The optimisation of model (7.2) over the range $0 < a_1 < 2$ and $0 < a_2 < 2$ using the signal $u_0(t)$ (7.11) gives the model

$$K_{4opt} = \frac{1}{(s^2 + a_{1opt}s + 1)(s^2 + a_{2opt}s + 1)} \tag{7.12}$$

$$a_{1opt} = a_{opt} = 1.999$$

which minimises the value $I_2(u_0)$, which was 11.692 [V^{2}s] before optimisation, to the value $min[I_2(u_0)] = 5.816$ [V^{2}s]. Diagrams in Fig.7.3 and Fig.7.4 present the responses of models $K_6(s)$, $K_4(s)$ and $K_{4opt}(s)$ and respective mapping errors in the case when the input signal $u_{0,4}(t)$ has been applied with switching instants given by (7.11).

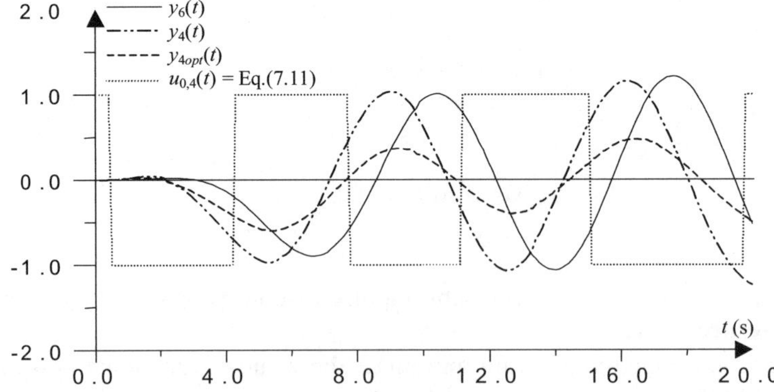

Fig. 7.3. Responses $y_6(t)$, $y_4(t)$ and $y_{4opt}(t)$ of the models $K_6(s)$, $K_4(s)$ and $K_{4opt}(s)$ to the signal $u_{0,4}(t)$ (7.11) .

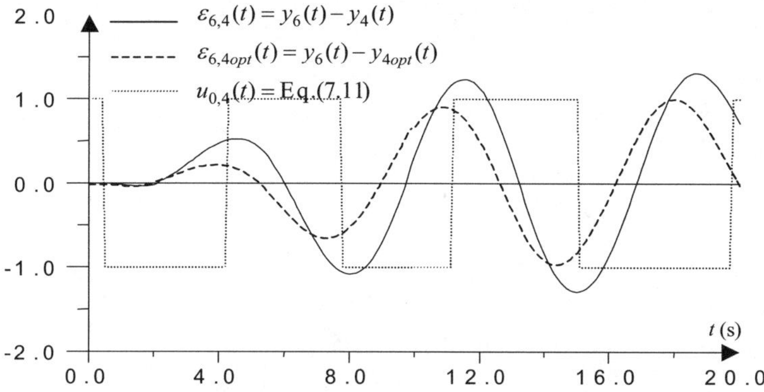

Fig. 7.4. Mapping errors of the model $K_6(t)$ by the model $K_4(t)$ before optimisation $\varepsilon_{6,4}(t) = y_6(t) - y_4(t)$ and after optimisation $\varepsilon_{6,4opt}(t) = y_6(t) - y_{4opt}(t)$ using the signal

$u_{0,4}(t)$ (7.11) .

Signal $u_0(t)$ with two constraints

The signal $u_0(t)$ with magnitude (7.7) and rate of change (7.8) constraints giving over [0,20s] the maximum value of 2.649 [V^2s] for the functional $I_2(u_0)$ is in the form

$$u_{04}(t) \Rightarrow \vartheta_+[0.00, 2.807],\ \vartheta_-[2.807, 7.570],\ \vartheta_+[7.570, 12.332],$$
$$\vartheta_-[12.332, 17.956],\ -1[17.956, 20.00] \tag{7.13}$$

where the tangent of the rise angle $\vartheta_+[.]$ is equal to the tangent of the fall angle $\vartheta_-[.]$ with a value of 0.355. The optimisation of model (7.2) using the signal $u_0(t)$ (7.13) gives the model

$$K_{4opt}(s) = \frac{1}{(s^2 + a_{1opt}s + 1)(s^2 + a_{2opt}s + 1)} \tag{7.14}$$

$$a_{1opt} = 1.998,\quad a_{21opt} = 1.478$$

which minimises the value $I_2(u_0)$, which was 2.649 [V^2s] before optimisation to the value $min[I_2(u_0)] = 1.086$ [V^2s]. The diagrams in Fig.7.5 and Fig.7.6 present the responses of models $K_6(s)$, $K_4(s)$ and $K_{4opt}(s)$ and their respective mapping errors in the case when input signal $u_0(t)$ (7.13) has been applied.

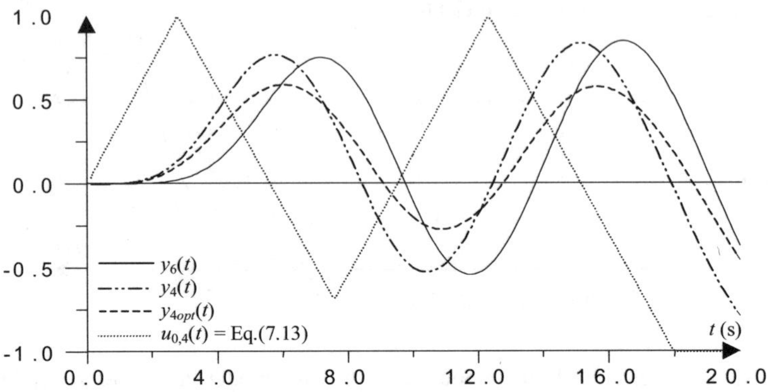

Fig. 7.5. Responses $y_6(t)$, $y_4(t)$ and $y_{4opt}(t)$ of the models $K_6(s)$, $K_4(s)$ and $K_{4opt}(s)$ to the signal $u_{0,4}(t)$ (7.13) .

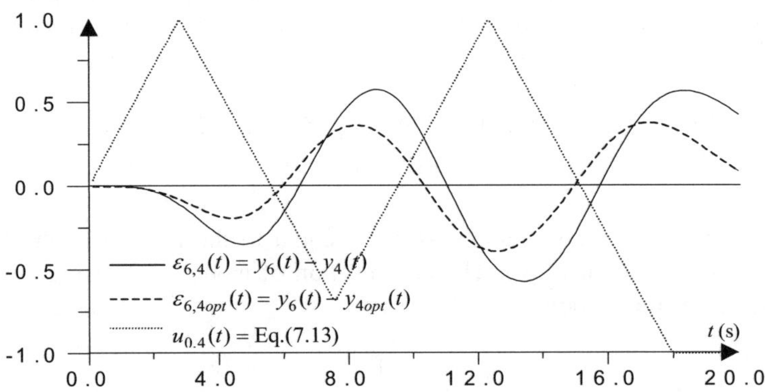

Fig. 7.6. Mapping errors of the model $K_6(t)$ by the model $K_4(t)$ before optimisation $\varepsilon_{6,4}(t) = y_6(t) - y_4(t)$ and after optimisation $\varepsilon_{6,4opt}(t) = y_6(t) - y_{4opt}(t)$ using the signal $u_{0,4}(t)$ (7.13) .

Optimisation of the third order filter for T = 20s.

Signal $u_0(t)$ with one constraint

Repeating an identical calculation procedure for the third order filter (7.3) as was applied for the fourth order model, the $u_0(t)$ signal of "bang-bang" type, which gives a maximum value of the functional $I_2(u_0)$ of 22.393 [V^{2}s] over [0,20s], has been obtained for the following switching instants

$$u_{03}(t) \Rightarrow t_0 = 0s., \quad t_1 = 0.519s., \quad t_2 = 4.402s., \quad t_3 = 8.013s.,$$
$$t_4 = 11.641s., \quad t_5 = 15.521s., \quad t_6 = T = 20.00s. \tag{7.15}$$

The optimisation of filter (7.3) using the signal $u_0(t)$ (7.15) gives the model

$$K_{3opt} = \frac{1}{(s^2 + a_{1opt}s + 1)(a_{2opt}s + 1)} \tag{7.16}$$
$$a_{1opt} = a_{2opt} = 1.999$$

which minimises the value $I_2(u_0)$, which was 22.393[V^{2}s] before optimisation to the value $min[I_2(u_0)] = 11.934[V^2s]$.

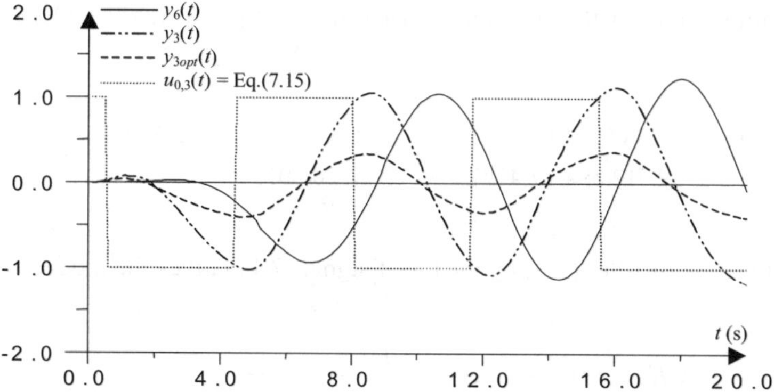

Fig. 7.7. Responses $y_6(t)$, $y_3(t)$ and $y_{3opt}(t)$ of the models $K_6(s)$, $K_3(s)$ and $K_{3opt}(s)$ to the signal $u_{0,3}(t)$ (7.15) .

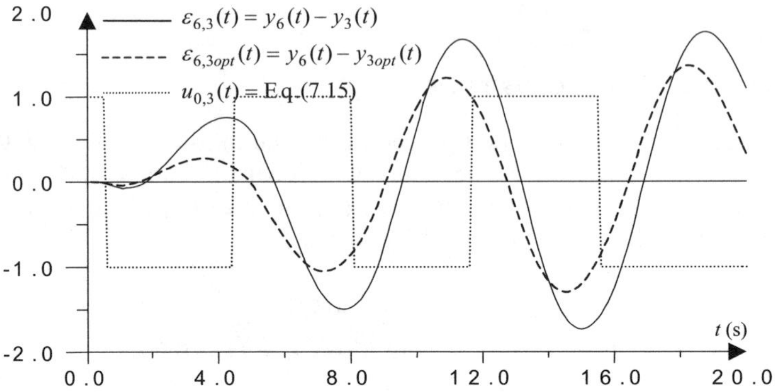

Fig. 7.8. Mapping errors of the model $K_6(t)$ by the model $K_3(t)$ before optimistion $\varepsilon_{6,3}(t) = y_6(t) - y_3(t)$ and after optimisation $\varepsilon_{6,3opt}(t) = y_6(t) - y_{3opt}(t)$ using the signal $u_{0,3}(t)$ (7.15) .

Signal $u_0(t)$ with two constraints

The signal $u_0(t)$ with magnitude (7.7) and rate of change (7.8) constraints giving the maximum value of the functional $I_2(u_0)$ of 5.498 [V²s] over [0,20s] is in the form

$$u_{03}(t) \Rightarrow \vartheta_+[0.00, 2.817], +1[2.817, 7.711], \vartheta_-[7.7711, 12.594],$$
$$\vartheta_+[12.594, 17.477], +1[17.477, 20.00]. \tag{7.17}$$

The optimisation of filter (7.3) by means of signal (7.17) gives the model

$$K_{3opt}(s) = \frac{1}{(s^2 + a_{1opt}s + 1)(a_{2opt}s + 1)} \tag{7.18}$$
$$a_{1opt} = 1.999, \quad a_{2opt} = 0.881$$

which minimises the value $I_2(u_0)$, which was 5.498 [V²s] before optimisation to the value $min[I_2(u_0)] = 1.539$ [V²s].

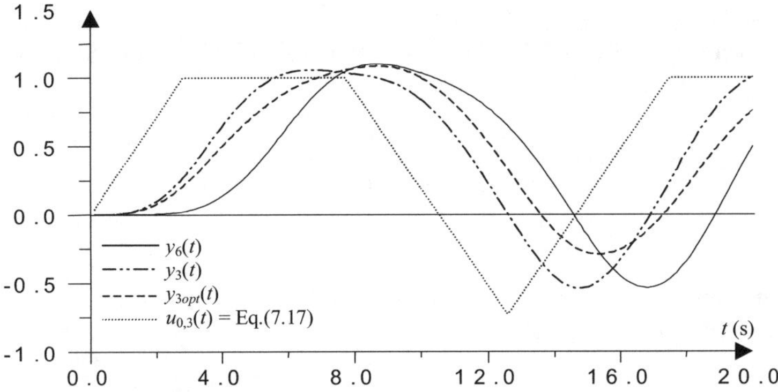

Fig. 7.9. Responses $y_6(t)$, $y_3(t)$ and $y_{3opt}(t)$ of the models $K_6(s)$, $K_3(s)$ and $K_{3opt}(s)$ to the signal $u_{0,3}(t)$ (7.17) .

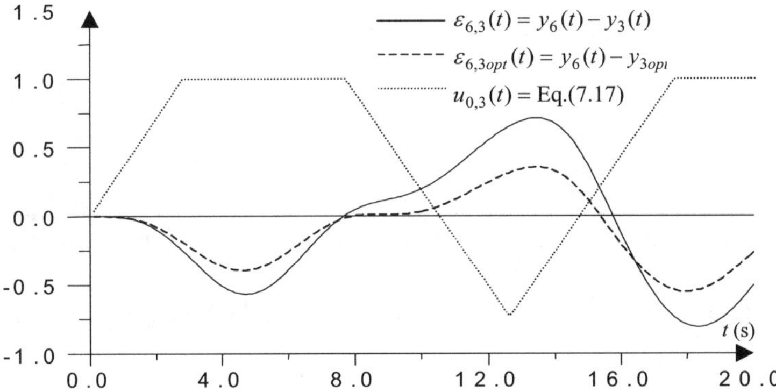

Fig. 7.10. Mapping errors of the model $K_6(t)$ by the model $K_3(t)$ before optimisation $\varepsilon_{6,3}(t) = y_6(t) - y_3(t)$ and after optimisation $\varepsilon_{6,3opt}(t) = y_6(t) - y_{3opt}(t)$ using the signal $u_{0,3}(t)$ (7.17) .

Optimisation of the fourth order filter for T = 10s.

Signal u₀(t) with one constraint

The signal $u_0(t)$, giving the maximum value of the functional $I_2(u_0)$ of 3.4767 [V²s] over [0,10s], has been obtained for the following switching instants

$$u_{04}(t) \Rightarrow t_0 = 0s., \; t_1 = 2.067s., \; t_2 = 5.652s., \; t_3 = T = 10.00s. \tag{7.19}$$

The optimisation of filter (7.2) using the signal $u_0(t)$ (7.19) gives the model

$$K_{4opt}(s) = \frac{1}{(s^2 + a_{1opt}s + 1)(s^2 + a_{2opt}s + 1)} \qquad (7.20)$$

$$a_{1opt} = 1.962, \quad a_{2opt} = 1.489$$

which minimises the value $I_2(u_0)$, which was 3.477 [V²s] before optimisation, to the value $min[I_2(u_0)] = 1.819[V^2s]$

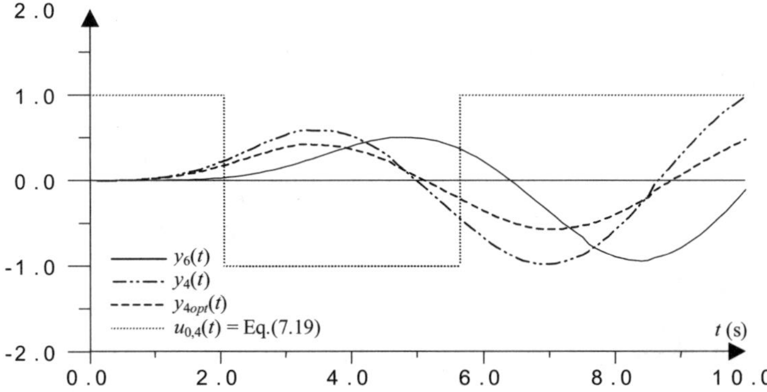

Fig. 7.11. Responses $y_6(t)$, $y_4(t)$ and $y_{4opt}(t)$ of the models $K_6(s)$, $K_4(s)$ and $K_{4opt}(s)$ to the signal $u_{0,4}(t)$ (7.19) .

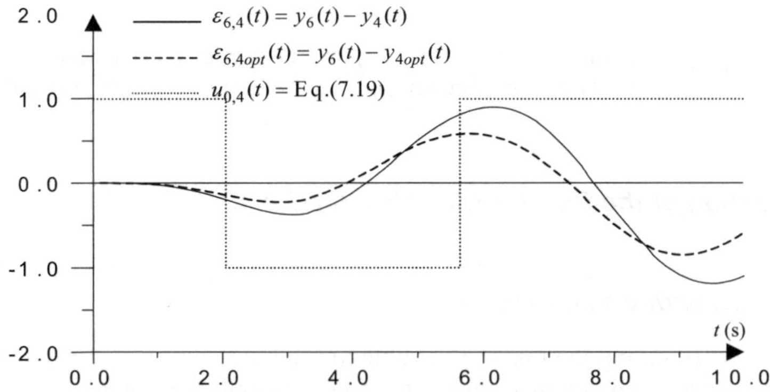

Fig. 7.12. Mapping errors of the model $K_6(t)$ by the model $K_4(t)$ before optimisation $\varepsilon_{6,4}(t) = y_6(t) - y_4(t)$ and after optimisation $\varepsilon_{6,4opt}(t) = y_6(t) - y_{4opt}(t)$ using the signal $u_{0,4}(t)$ (7.19).

Signal $u_0(t)$ with two constraints

The signal $u_0(t)$ with constraints (7.7) and (7.8) giving the maximum value of the functional $I_2(u_0)$ of 0.991 $[V^2s]$ over $[0,10s]$ is in the form

$$u_{04}(t) \Rightarrow \vartheta_-[0.00,2.817],\ \vartheta_+[2.817,8.449],\ +1[8.449,10.00]\ . \qquad (7.21)$$

The optimisation of filter (7.2) using the signal $u_0(t)$ (7.21) the gives model

$$K_{4opt}(s) = \frac{1}{(s^2 + a_{1opt}s +1)(s^2 + a_{2opt}s +1)} \qquad (7.22)$$

$$a_{1opt} = a_{2opt} = 1.999$$

which minimises the value of $I_2(u_0)$, which was $0.991[V^2s]$ before optimisation to the value $min[I_2(u_0)] = 0.299\ [V^2s]$

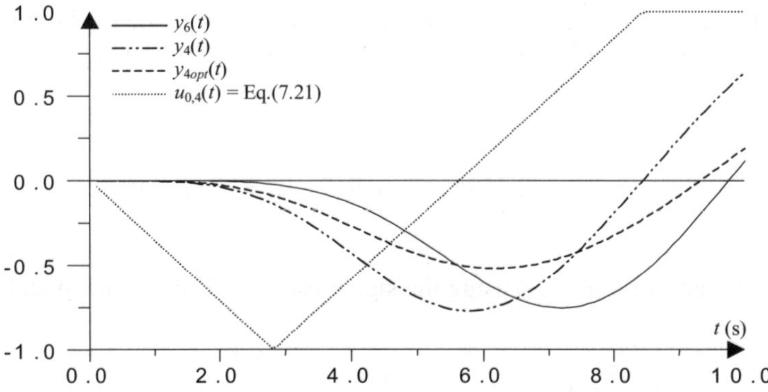

Fig. 7.13. Responses $y_6(t)$, $y_4(t)$ and $y_{4opt}(t)$ of the models $K_6(s)$, $K_4(s)$ and $K_{4opt}(s)$ to the signal $u_{0,4}(t)$ (7.21) .

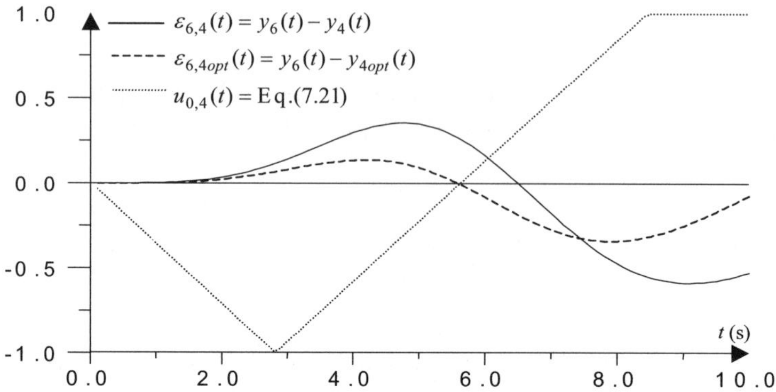

Fig. 7.14. Mapping errors of the model $K_6(t)$ by the model $K_4(t)$ before optimisation $\varepsilon_{6,4}(t) = y_6(t) - y_4(t)$ and after optimisation $\varepsilon_{6,4opt}(t) = y_6(t) - y_{4opt}(t)$ using the signal $u_{0,4}(t)$ (7.21) .

Optimisation of the third order filter for T = 10s.

Signal $u_0(t)$ with one constraint

The signal $u_0(t)$ of "bang-bang" type giving the maximum value of the functional $I_2(u_0)$ of 7.472 [V^2s] over [0,10s] has been obtained for the following switching instants

$$u_{03}(t) \Rightarrow t_0 = 0s., \quad t_1 = 2.358s., \quad t_2 = 6.017s., \quad t_3 = T = 10.00s. \tag{7.23}$$

The optimisation of filter (7.3) using the signal $u_0(t)$ (7.23) gives the model

$$K_{3opt}(s) = \frac{1}{(s^2 + a_{1opt}s + 1)(a_{2opt}s + 1)} \tag{7.24}$$

$$a_{1opt} = 1.999, \quad a_{2opt} = 0.186$$

which minimises the value of $I_2(u_0)$, which was 7.472[V^2s] before optimisation to the value $min[I_2(u_0)] = 2.946$ [V^2s].

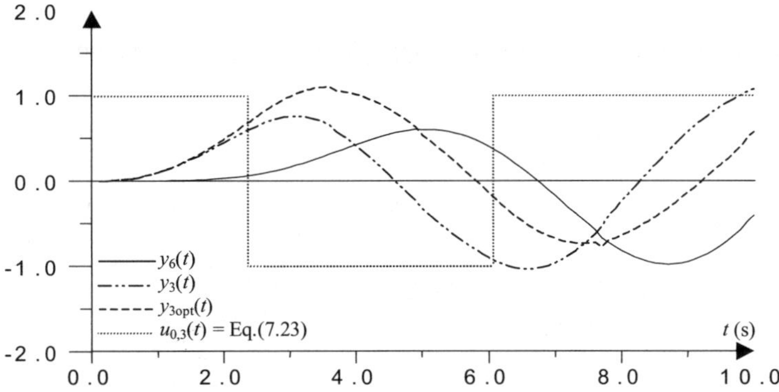

Fig. 7.15. Responses $y_6(t)$, $y_3(t)$ and $y_{3opt}(t)$ of the models $K_6(s)$, $K_3(s)$ and $K_{3opt}(s)$ to the signal $u_{0,3}(t)$ (7.23) .

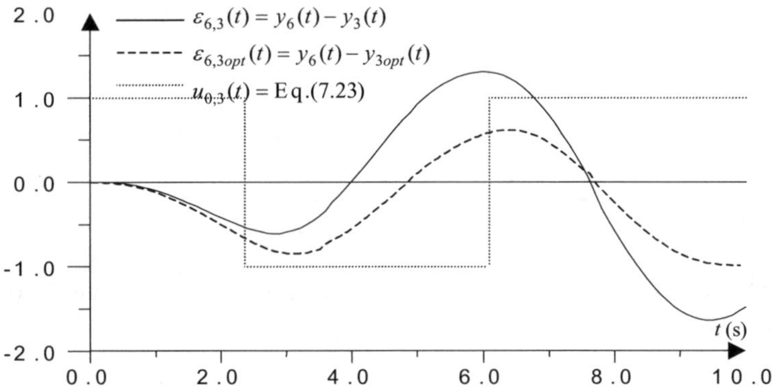

Fig. 7.16. Fig.7.16. Mapping errors of the model $K_6(t)$ by the model $K_3(t)$ before optimisation $\varepsilon_{6,3}(t) = y_6(t) - y_3(t)$ and after optimisation $\varepsilon_{6,3opt}(t) = y_6(t) - y_{3opt}(t)$ using the signal $u_{0,3}(t)$ (7.23) .

Signal $u_0(t)$ with two constraints

The signal $u_0(t)$ with constraints (7.7) and (7.8) giving the maximum value of the functional $I_2(u_0)$ of 2.199 [V^{2}s] over [0,10s] has an identical form to that of signal (7.21)

$$u_{03}(t) \Rightarrow \vartheta_-[0.00, 2.817], \vartheta_+[2.817, 8.449], \vartheta_-[8.449, 10.00] . \tag{7.25}$$

The optimisation of filter (7.3) by means of signal (7.25) gives the model

$$K_{3opt}(s) = \frac{1}{(s^2 + a_{1opt}s + 1)(a_{2opt}s + 1)}$$

$$a_{1opt} = 1.999 \quad a_{2opt} = 0.658$$

(7.26)

which minimises the value $I_2(u_0)$, which was 2.199 [V^{2}s] before optimisation to the value $min[I_2(u_0)] = 0.858$ [V^{2}s]

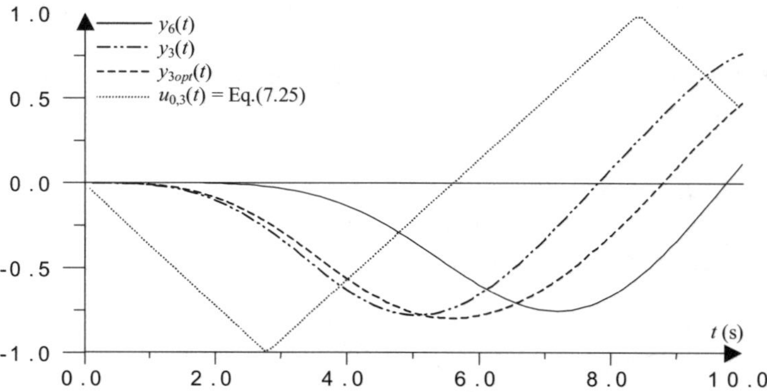

Fig. 7.17. Responses $y_6(t)$, $y_3(t)$ and $y_{3opt}(t)$ of the models K_6(s), K_3(s) and K_{3opt}(s) to the signal $u_{0,3}(t)$ (7.25) .

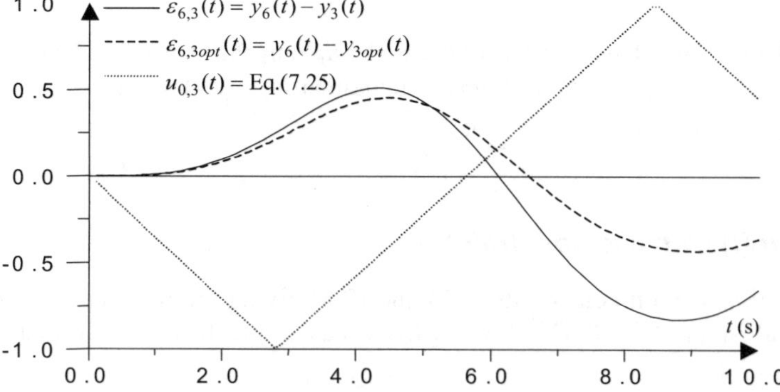

Fig. 7.18. Mapping errors of the model K_6(s) by the model K_3(s) before optimisation $\varepsilon_{6,3}(t) = y_6(t) - y_3(t)$ and after optimisation $\varepsilon_{6,3opt}(t) = y_6(t) - y_{3opt}(t)$ using the signal $u_{0,3}(t)$ (7.25) .

As a result of the optimisation carried out by means of the fourth and third order optimum models, the primary mapping errors of the sixth order model have
been reduced from 1.9 times for model (7.16) and signal (7.15) to 3.6 times for
model (7.18) and signal (7.17). Better results of optimisation were impossible to
attain due to essential constraints concerning the permissible adjustment range of
the models coefficients a_i. It is worth to emphasising that all the optimum models
have properties similar to those of Bessel filters. All of them determined for signals with one as well as with two constraints do not exhibit any oscillations in step
responses and exhibit almost linear phase-frequency characteristics. The highest
non-linearity does not exceed a few percentage points in any considered case.

7.3. Optimisation of models in the case of the small value of primary mapping error

The results of optimisation presented in Example 7.2 refer to models indicating a
high value of mapping errors before optimisation. On the other hand the process of
optimisation has been carried out over a very limited range, as a result of the permissible values of the tuned coefficients. Below we will carry out an optimisation
of the second order model, mapping the sixth order model in the case in which the
primary mapping error already has a very small value. For this purpose we will
use the models presented in examples 6.1 and 6.2.

7.4. Examples

Example 7.2

Determine the optimum second order model (6.61) with respect to *minimax* $I_2(u_0)$,
referring to the sixth order model (6.60).

Solution

In the sequel we will use the results of calculations of the $u_0(t)$ signals presented in
examples 6.1 and 6.2. Hence we have the sixth order model (6.60)

$$K_6(s) = \frac{6s^4 + 50s^3 + 196s^2 + 418s + 434}{s^6 + 12s^5 + 71s^4 + 256s^3 + 575s^2 + 804s + 585} \tag{7.27}$$

and mapping it, the second order model (6.61)

$$K_2(s) = \frac{b_0}{s^2 + a_1 s + a_0} \tag{7.28}$$

$$a_1 = 3.66, \quad a_0 = 7.78, \quad b_0 = 6.00.$$

7.2.1. Results of optimisation

Signal $u_0(t)$ with one constraint

Optimisation of model (7.28) by means of signal $u_{0,8}(t)$ (6.71) for $T = 8$s gives the model

$$K_{2opt}(s) = \frac{b_{0opt}}{s^2 + a_{1opt}s + a_{0opt}}$$

(7.29)

$$a_{1opt} = 3.220 \quad a_{0opt} = 7.129 \quad b_{0opt} = 5.538$$

which minimises the value $I_2(u_0)$, which was $14.71 \cdot 10^{-3}$ [V^{2}s] before optimisation to the value $min[I_2(u_0)] = 1.99 \cdot 10^{-3}$ [V^{2}s].

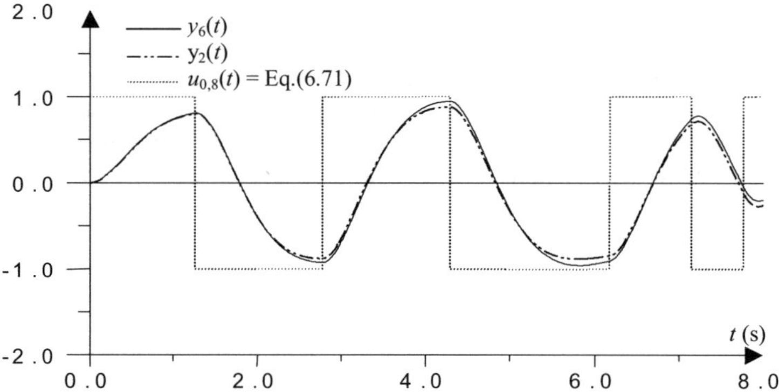

Fig. 7.19. Responses $y_6(t)$ and $y_2(t)$ of the models $K_6(s)$ and $K_2(s)$ to the signal $u_{0,8}(t)$ (6.71) .

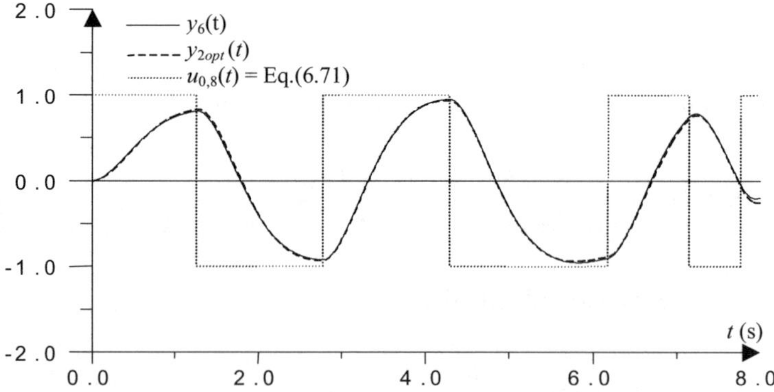

Fig. 7.20. Responses $y_6(t)$ and $y_{2opt}(t)$ of the models $K_6(s)$ and $K_{2opt}(s)$ to the signal $u_{0,8}(t)$ (6.71) .

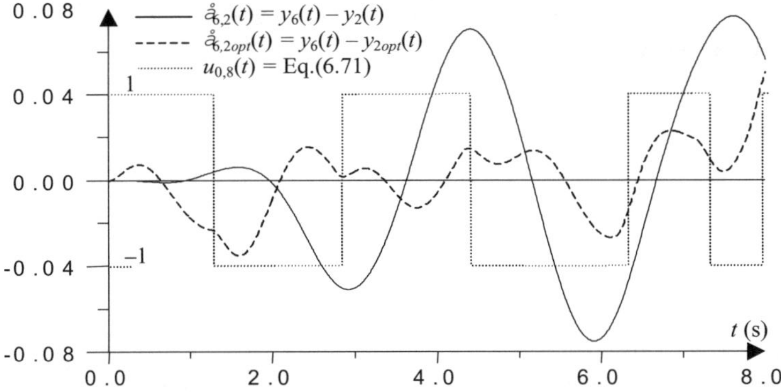

Fig. 7.21. Mapping errors of the model $K_6(s)$ by the models $K_2(s)$ and $K_{2opt}(s)$ before optimisation $\varepsilon_{6,2}(t) = y_6(t) - y_2(t)$ and after optimisation $\varepsilon_{6,2opt}(t) = y_6(t) - y_{2opt}(t)$ using the signal $u_{0,8}(t)$ (6.71)

Signal $u_0(t)$ with two constraints

Optimisation of model (7.28) using the signal $u_{0,3}(t)$ (6.76), where $\vartheta = 1.02$ (6.73), gives the model

$$K_{2opt}(s) = \frac{b_{0opt}}{s^2 + a_{1opt}s + a_{0opt}} \tag{7.30}$$

$$a_{1opt} = 3.040 \quad a_{0opt} = 7.419 \quad b_{0opt} = 5.492$$

which minimises the value $I_2(u_0)$, which was $58.43 \cdot 10^{-4}$ [V^{2}s] before optimisation to the value $min[I_2(u_0)] = 3.14 \cdot 10^{-4}$ [V^{2}s].

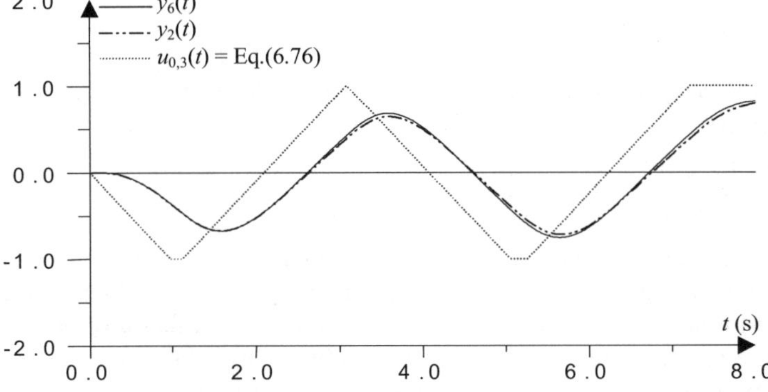

Fig. 7.22. Responses $y_6(t)$ and $y_2(t)$, of the models $K_6(s)$ and $K_2(s)$ to the signal $u_{0,3}(t)$ (6.76) .

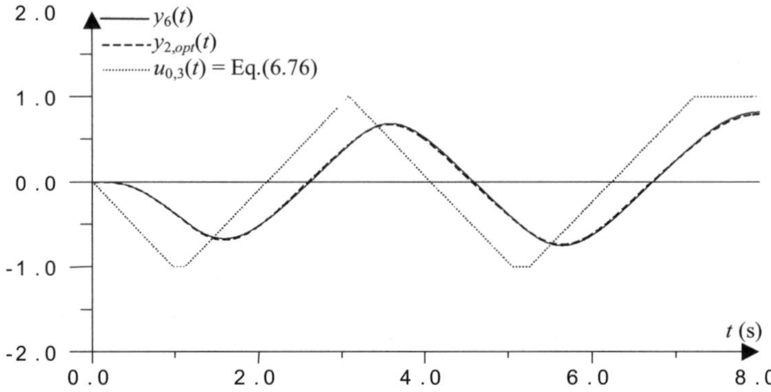

Fig. 7.23. Responses $y_6(t)$ and $y_{2opt}(t)$ of the models $K_6(s)$ and $K_{2opt}(s)$ to the signal $u_{0,3}(t)$ (6.76) .

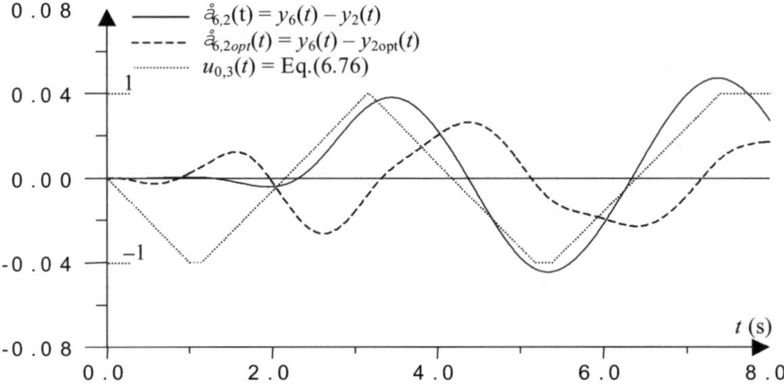

Fig. 7.24. Mapping errors of the model $K_6(s)$ by the model $K_2(s)$ and $K_{2opt}(s)$ before optimisation $\varepsilon_{6,2}(t) = y_6(t) - y_2(t)$ and after optimisation $\varepsilon_{6,2opt}(t) = y_6(t) - y_{2opt}(t)$ using the signal $u_{0,3}(t)$ (6.76) .

The example presented above shows that the parameters of the optimum model for a signal of "bang-bang" type reduce the mapping error more than by a factor of seven, and for signals with two constraints more than by a factor of eighteen. This points to the fact that the optimisation method by means of signals maximising the error criterion, is featured by a high sensitivity, both with respect to the shape of the input signal as well as to the parameter values of the optimised model. It provides a good outlook on conducting effective optimisation of different systems when strictly determined constraints, resulting from their dynamic properties, are imposed on the input signals. The method presented here seems to be particularly useful for determining optimum models of systems for which, as originally assumed, the input signals are unknown and varying in time. In the case of such systems the problem arises of how to choose the most appropriate input signal. The application of signals, which maximise the error criterion, solves this problem to a significant degree. These allow us to determine optimum models, invariant to shapes of dynamic signals, in such a sense that the mapping error for these models is valid for any signals which could appear at the input of the real system. Presented examples refer to the integral-square-error criterion, but can also be easily extended to other criteria. It is worth to emphasising here that the application of programs based on genetic algorithms allows the optimisation results to be obtained quickly and easily for different objective functions, and particularly in difficult cases where several constraints are imposed on the input signals simultaneously.

REFERENCES

[1] Ahlberg JH, Nilson EN and Walsh JL (1967) The Theory of Splines and Their Applications. Academic Press, Mathematics in Science and Engineering, New York and London

[2] Aldhaheri RW (1991) Model reduction via Shur-form decomposition. Int.J. Control 53: 709-716

[3] Allemandou P (1966) Low-pass filters approximating - in modulus and phase - the exponential function. IEEE Transactions on Circuit Theory 13: 298-301,

[4] Al-Saggaf UM, Franklin GE (1988) Model Reduction Via Balanced Realizations: An Extension and Frequency Weighting Techniques. IEEE Trans. Autom. Control 33: 687-691

[5] Arumugam M, Ramamoorty M. (1973) A method of simplifying large dynamic systems. Int. J. Control 17: 1129-1135

[6] Barnett S (1973) New reductions of Hurwitz determinants. Int. J. Control 18: 977-991

[7] Bendat JS, Piersol AG (1976) Metody analizy i pomiaru sygnalow losowych. PWN, Warszawa

[8] Birch BJ, Jackson R (1959) The behaviour of linear systems with inputs satisfying certain bounding conditions. J. Electronics and Control 6: 366-375

[9] Bosley MJ, Kropholler HW and Lees FP (1973) On the relation between the continued fraction expansion and moments matching methods of model reduction. Int. J. Control 18: 461-474

[10] Burden RL, Faires JD (1985) Numerical Analysis. PWS-KENT Publishing Company, Boston

[11] Candy JV (1988) Signal Processing. The Modern Approach. McGrow-Hill Book Company, New York

[12] Chen CF, Chu H (1966) A Matrix for Evaluating Schwarz's Form. IEEE Trans. Autom. Control 4: 303-305,

[13] Chen CF, Shieh LS (1968) A novel approach to linear model simplification. Int. J. Control 8: 561-570

[14] Ching-Tien L, Yi-Shyong C (1988) Successive parameter estimation of continuous dynamic systems. Int. J. Systems Science 19: 1149-1158

[15] Davison E J (1966) A Method for Simplification Linear Dynamic Systems. IEEE Trans. Autom. Control 11: 93-101

[16] Director SW, Rother RA (1972) Introduction To Systems Theory. McGraw-Hill, New York

[17] Eykhoff P (1974) System Identification. Parameter and State Estimation. John Wiley and Sons. London

[18] Fhrumann A (1996) A Polynomial Approach to Linear Algebra. Springer-Verlag, New York

[19] Friedland B (1989) On the Properties of Reduced-Order Kalman Filters. IEEE Trans. Autom. Control 34: 321-324

[20] Fuksa S, Byrski W (1980) Problem optymalizacji pewnego typu funkcjonalow kwadratowych na zbiorach wypuklych. Prace VIII Krajowej Konferencji Automatyki. Szczecin: 62-64

[21] Gajda J, Szyper M. (1998) Modelowanie i badania symulacyjne systemow pomiarowych. Jartek s.c. Krakow

[22] Gawransky W, Natke HG (1988) Order estimation of AR and APMA models. Int. J. Systems Science 19: 1143-1148

[23] Goldberg DE (1989) Genetic Algorithms in Search, Optimization, and Machine Learning. Addison-Wesley Publishing Company, USA

[24] Gutman P, Mannerfelt F and Molamder P (1982) Contribution to the Model Reduction Problem. IEEE Trans. Autom. Control 27: 454-455,

[25] Hagel R, Zakrzewski J (1984) Miernictwo dynamiczne. WNT, Warszawa

[26] Halevi Y (1996) Reduced order models with delay. Int. J. Control 64: 733-744,

[27] Helavi Y (1992) Frequency Weighted Model Reduction via Optimal Projection. IEEE Trans. Autom. Control 37: 1537-1542

[28] Hutton MF, Friedland B (1975) Routh approximations for reducing order of linear time-invariant systems. IEEE Trans. Autom. Control 20: 329-337

[29] Hwang C, Lee Y (1989) Multifrequency Padé Approximation Via Jordan Continued-Fraction Expansion. IEEE Trans. Autom. Control 34: 444-446

[30] Johnson CD, Wonham WM (1966) Another Note on the Transformation to Canonical (Phase-Variable) Form. IEEE Trans. on Autom. Control. 7: 609-610

[31] Kaczorek T (1996) Teoria sterowania i systemow. PWN, Warszawa

[32] Kong SY, Lin DW (1981) Optimal Henkel Norm Model Reductions: Multivariable Systems. IEEE Trans. Autom. Control 26: 832-852

[33] Kordylewski W, Wach J (1988) Usrednione rozniczkowanie zakloconych sygnalow pomiarowych. PAK 6: 123-124

[34] Krishnamurthy V, Seshadri V (1976) A simple and direct method of reducing the order of linear, time-invariant systems by Routh approximation frequency domain. IEEE Trans. Autom. Control. 20: 797-799

[35] Krishnamurthy V, Seshadri V (1978) Model Reduction Using the Routh Stability Criterion. IEEE Trans. Autom. Control 23: 729-731

[36] Ku YH (1961) Transient Circuit Analysis. D. Van Nostrand Co. Inc., Princeton

[37] Lam J (1993) Model reduction of delay systems using Padé approximation. Int. J. Control 57: 377-391

[38] Lam J, Yang G (1996) Balanced model reduction of symmetric composite systems. Int. J. Control 65: 1031-1043,

[39] Lamba S, Goerz R and Bandyopadhyay B (1988) New reduction technique by step error minimization for multivariable systems. Int. J. Systems Science 19: 999-1009

[40] Layer E (1981) Theoretical foundations of the calibration process of measuring systems in the aspect of dynamic errors. Proc. IMEKO Symposium on Computerized Measurement, Dubrovnik: 113-116

[41] Layer E (1981) Podstawy teorii wzorcowania systemow pomiarowych w aspekcie bledow dynamicznych. ZN AGH, Krakow

[42] Layer E (1982) Basic problems of the calibration process and of the establishing a hierarchy of accuracy for dynamic measuring systems. Proc. IMEKO 9th World Congress, Berlin (West) 5/3: 269-277

[43] Layer E (1997) Theoretical Principles for Establishing a Hierarchy of Dynamic Accuracy with the Integral-Square-Error as an Example. IEEE Trans. Instrumentation and Measurement 46: 1178-182

[44] Layer E (1999) Mapping Error of Simplified Dynamic Models in Electrical Metrology. Proc. 16th IEEE Instrumentation and Measurement Technology Conference, Venice 3: 1704-1709

[45] Layer E (2001) Mapping Error of Linear Dynamic Systems Caused by Reduced-Order Model. IEEE Trans. Instrumentation and Measurement 50: 792-800

[46] Layer E (2001) Przestrzen rozwiazan sygnalow maksymalizujacych kwadratowy wskaznik jakosci. Mat. Konf. Modelowanie i symulacja systemow pomiarowych. AGH, Krynica: 25-28

[47] Layer E, Gawedzki W (1991) Dynamika aparatury pomiarowej. Badania i ocena. PWN, Warszawa

[48] Layer E, Gawedzki W (1993) Time frequency properties of signals maximizing the dynamic errors. Systems-Analysis-Modelling-Simulation, Gordon & Breach Science Publisher 11: 73-77,

[49] Layer E, Piwowarczyk T (1999) Application of the generalized Fibonacci sequences to the simplification of mathematical models of linear dynamic systems. Archives of Electrical Engineering. Polish Scientific Publishers PWN, Warsaw 187/188: 19-30

[50] Layer E, Piwowarczyk T (2000) Generalised Fibonacci Series in the Description of Dynamic Models. Systems-Analysis-Modelling-Simulation, Gordon & Breach Science Publisher 37: 57-67

[51] Leitner R (1995) Zarys matematyki wyzszej. WNT, Warszawa

[52] Ljubojevicz M (1973) Suboptimal input signal for linear system identification. Int. J. Control 17: 659-669

[53] Luenberger DG (1967) Cannonical forms for linear multivariable systems. IEEE Trans. Autom. Control 12: 290-293

[54] Luke YL (1969) The Special Functions and their Approximations. Academic Press, New York

[55] Luke YL (1975) Mathematical Functions and their Approximations. Academic Press, New York,

[56] Marczuk GI (1983) Analiza numeryczna zagadnien fizyki matematycznej. PWN Warszawa

[57] Meier L, Luenberger DG (1967) Approximation of linear constant systems. IEEE Trans. Autom. Control 12: 585-588

[58] Moore BC (1981) Principal Component Analysis in Linear Systems: Controllability, Observability, and Model Reduction. IEEE Trans. Autom. Control 26: 17-31,

[59] Noble B, Daniel JW(1997) Applied Linear Algebra. Prentice-Hall Inc. Englewood Cliffs, New Jersey

[60] Orlowski M. (1992) Odtwarzanie usrednionych sygnalow wejsciowych na podstawie zaszumionych sygnalow wyjsciowych. Ph.D Thesis, Politechnika Szczecinska, Szczecin

[61] Pernebo L, Silverman LM (1982) Model reduction via balanced-truncation model reduction. IEEE Trans. Autom. Control 27: 382-387

[62] Piotrowski J, Kostyrko K (2000) Wzorcowanie aparatury pomiarowej. PWN Warszawa

[63] Piwowarczyk T (1996) Coefficients of power expansion of original as function of transform coefficients. Cracow University of Technology, Czasopismo Techniczne, 93/7 E: 21-30

[64] Piwowarczyk T (2000) Multipower Notation of Symmetric Polynomials in Engineering Calculus. Wydawnictwo Instytutu Gospodarki Surowcami Mineralnymi i Energia PAN, Krakow

[65] Rane DS (1996) A Simple Transformation to (Phase-Variable) Canonical Form. IEEE Trans. Autom. Control 7: 608

[66] Rao S.A., Lamba SS (1974) A new frequency domain technique for the simplification of linear dynamic systems. Int. J. Control 20: 71-79

[67] Rutland NK (1994) The Principle of Matching: Practical Conditions for Systems with Inputs Restricted in Magnitude and Rate of Change. IEEE Trans. Autom. Control 39: 550-553

[68] Safonov MG, Chiang RY (1989) A Schur Method for Balanced-Truncation Model Reduction". IEEE Trans. Autom. Control 34/7: 729-733

[69] Sarma IG, Pai MA and Viswanathan R (1968) On the Transformation to Schwarz Canonical Form. IEEE Trans. Autom. Control 6: 311-312

[70] Shamash Y (1975) Model reduction using Routh stability criterion and the Pade approximation technique. Int. J. Control 21: 475-484

[71] Sinha NK, Bereznai GT (1971) Optimum approximation of high-order systems by low order model. Int. J. Control 21: 951-959

[72] Sinha NK, De Bruin H (1973) Near optimal control of high-order systems using low-order models. Int. J. Control 17: 257-262

[73] Sinha NK, Pille W (1971) A new method for reduction of dynamic systems. Int. J. Control 14: 111-118

[74] Skoczowski S (1995) Identyfikacja praktyczna modeli transmitancyjnych na podstawie poczatkowej fazy odpowiedzi skokowej. Kwartalnik Elektroniki i Telekomunikacji 41: 5-29

[75] Tietze U, Schenk C (1974) Halbleiter- Schaltungstechnik. Springer-Verlag, Berlin

[76] Tsu-Tian L, Shuh-Chuan T (1988) Solution of linear dynamic systems with initial or boundry value conditions by shifted Chebyshev approximations. Int. J. Systems Sciences 19: 1225-1234

[77] Tuel WG Jr. (1966) On the Transformation to (Phase-Variable) Canonical Form. IEEE Trans. Autom. Control 7: 607

[78] Whitefield AH, Williams NG (1988) Integral least-squares techniques for frequency-domain model reduction. Int. J. Systems Science 19: 1355-1373

[79] Wilson DA (1974) Optimum solution of model reduction problem. Int. J. Control 20: 57-64

[80] Yasuyuki F, Kaheu N (1973) Identification of discrete-time systems using a composite sinusoidal signal. Int. J. Control 18: 945-962

[81] Zakian V (1973) Simplification of linear time-invariant systems by moment approximants. Int. J. Control 18: 455-460

[82] Zakian V (1989) Critical systems and tolerable inputs. Int. J. Control 49: 1285-1289

[83] Zakian V (1996) Perspectives of the principle of matching and the method of inequalities. Int. J. Control 65: 147-175

[84] Zill DG (1986) Differential Equations with Boundry-Value Problems. Prindle, Weber & Schmidt, Boston

[85] Zakowski W, Leksinski W (1995) Matematyka, Czesc IV. WNT, Warszawa

[86] Zuchowski A (1998) Uproszczone modele dynamiki. Politechnika Szczecin ska, Instytut Automatyki Przemys owej. Szczecin

INDEX

Printing (Computer to Film): Saladruck Berlin
Binding: Stürtz AG, Würzburg